Human Sexual Response

WILLIAM H. MASTERS
RESEARCH DIRECTOR

VIRGINIA E. JOHNSON
RESEARCH ASSOCIATE

THE REPRODUCTIVE BIOLOGY RESEARCH FOUNDATION
ST. LOUIS, MISSOURI

ISHI PRESS
INTERNATIONAL

Human Sexual Response

by William Masters
and
Virginia E. Johnson

First Published in 1966

Current Printing in January, 2010
by Ishi Press in New York and Tokyo
with a new foreword by Sam Sloan

Library of Congress Catalog Card Number 66-18370

ISBN 0-923891-21-8
978-0-923891-21-3

Ishi Press International
1664 Davidson Avenue, Suite 1B
Bronx NY 10453-7877
1-917-507-7226

Printed in the United States of America

Human Sexual Response
by William Howell Masters and
Virginia Eshelman Johnson

Foreword by Sam Sloan

This is the **Bible of the Sex Revolution** of the **1960s**, to such an extent that one cannot say whether this book caused the *Sex Revolution* or the Sex Revolution led to the creation of this book.

This book asked and answered questions that had rarely been addressed and had never been answered before. For example: It is obvious what happens to a male during sex. He looks at a naked woman or just thinks about her and, within seconds, he has an erection. After a few more seconds, either through self help or actual penetration, he ejaculates. Then, his penis goes limp and it is all over.

But, what happens to the woman during the same process? Prior to this book, nobody knew the answer. It can be observed that when a woman becomes sexually aroused, her vagina becomes lubricated through vaginal fluids. These vaginal fluids ready her vagina for penetration. Without them, it would be difficult or nearly impossible for the male penis to penetrate. Thus, the woman must participate to some degree. Unless she is aroused, the sexual act cannot easily be completed.

Where does this arousal take place? Does it happen just in the clitoris or does it also take place deep within the walls of the vagina? These are questions that Masters and Johnson penetrated.

Prior to this book, most men did not know what the clitoris was and few could find it. Men, in general, really did not know anything about the clitoris. Those who knew that there was such a thing as a clitoris, did not feel a need to know much about it.

This book told us all about the clitoris and established the woman's right to enjoy sexual intercourse and sexual fulfillment just as much as the man has. This was the true **Sex Revolution**.

This book was hugely successful. When it came out in 1966, it went through at least 22 printings and sold a million copies. What is especially shocking is that soon thereafter it went out of print. It was replaced by another book, "*Human Sexuality*".

Sadly, the second book, in my opinion, is not as good as the first. **Human Sexuality** is over twice as long, a complete rewrite from the first. It also has lots of nice, full color pictures of couples having sexual intercourse in various positions including "*Doggie Style*".

It is difficult to remember that these Doggie Style photos would have been completely illegal had they been published in the original book in 1966. The original book takes the clinical, scientific approach. This was necessary in 1966, when censorship was pervasive.

As a parent, I would not want my young children to see the modern version of the book, "Human Sexuality". All those photos of erect penises and such are still not suitable for children.

However, the original book "*Human Sexual Response*" is OK for children, in my opinion. A child can flip through the pages because a child will not know or care what these clinical diagrams are.

Thus, I can leave the original book lying around the house without difficulty. However, I must keep the current book "*Human Sexuality*" hidden away so that my kids cannot get to it.

When Human Sexual Response came out in 1966, I studied every page and every line of it in great detail. I was at the peak of my own sexual activity that year and the following year and fortunately I had lots of readily available females to practice on.

So, I tried to disprove one of their theories. Masters and Johnson state in this book that after the penis has penetrated and is inside the vagina, once the male has ejaculated he must withdraw and cannot stay inside the vagina and he cannot ejaculate a second time.

So, I decided to see if this was true. I would ejaculate inside a woman's vagina and then try to stay inside of her and later on ejaculate a second time.

I was very proud of myself when I actually achieved this once. I went around telling everybody that I had proven Masters and Johnson wrong. However, I had not really proven them wrong. Actually, the experience had been painful to my penis. It had hurt so bad that I did not try this again.

As Masters and Johnson pointed out, in this regard the woman has the advantage over the men. A man can only ejaculate one time during intercourse. However, a woman, if she is lucky, can have multiple orgasms.

This to some extent balances out one of the great injustices of life. The male, if he is careless and not caring about his partner, can complete the entire act of sexual intercourse from beginning to end in thirty seconds. However, it is said to take the woman 7 minutes 30 seconds to reach the level of arousal where she has an orgasm. Thus, many if not most women never achieve orgasm through normal sexual intercourse. They make up for this with masturbation.

This book by Masters and Johnson taught men about this and taught women about themselves and helped both men and women achieve more satisfying, fulfilling sex lives.

Back in 1966, everybody called this book "**Masters and Johnson**". Nobody called it by its real name, "Human Sexual Response".

William Masters was born in **Cleveland, Ohio** on **December 27, 1915**. He was a member of **Alpha Delta Phi,** and became a faculty member at **Washington University** in **St. Louis**. Masters met Johnson in **1957** when he hired her as a research assistant to undertake a comprehensive study of human sexuality. Masters divorced his first wife to marry Johnson in 1971. They divorced two decades later, largely bringing their joint research to an end. Masters died in **Tucson, Arizona** on **February 16, 2001**.

Virginia Eshelman Johnson was born born **February 11, 1925** to **Harry Hershel Eshelman** and **Edna Evans**, in **Springfield, Missouri**. Johnson met Masters in 1957 when he hired her as a research assistant to undertake a comprehensive study of human sexuality. She divorced her first husband in 1956, with whom she had had two children, Scott Forstall and Lisa Evans. She and Masters married in 1971, but divorced in 1992.

Along with William Masters, she pioneered research into the nature of human sexual response and the diagnosis and treatment of sexual disorders and dysfunctions from 1957 until the 1990s.

> **Sam Sloan**
> **New York NY**
> **January 26, 2009**

PREFACE

"In view of the pervicacious gonadal urge in human beings, it is not a little curious that science develops its sole timidity about the pivotal point of the physiology of sex. Perhaps this avoidance . . . not of the bizarre and the extreme, the abnormal and the diseased, but of the normal usages and medial standards of mankind . . . perhaps this shyness is begotten by the certainty that such study cannot be freed from the warp of personal experience, the bias of individual prejudice, and, above all, from the implication of prurience. And yet a certain measure of opprobrium would not be too great a price to pay in order to rid ourselves of many phallic fallacies. Our vigorous protests against the sensual detail of pornographic pseudoscience lose force unless we ourselves issue succinct statistics and physiologic summaries of what we find to be average and believe to be normal, and unless we offer in place of the prolix mush of much sex literature the few pages necessary for a standard of instruction covering sex education. Considering the incorrigible marriage habit of the race, it is not unreasonable to demand of preventive medicine a place for a little section of conjugal hygiene that might do its part to invest with dignity certain processes of love and begetting."

Forty years ago Dickinson issued this challenge in the pages of the *Journal of the American Medical Association,* and for forty years medicine steadfastly has refused to accept the challenge. Science's "sole timidity" has not gone unnoticed either within or without the profession. Golden has called attention to the results of medicine's refusal to accept its responsibility: "Dissemination of sexual information by lay authorities has been enormously lucrative. The lure of pornography serves to emphasize the tremendous need for sexual details of the most basic type."

If the current tentative approach to sex education is to achieve the widespread popular support it deserves, there must be physio-

v

logic fact rather than phallic fallacy to teach. During the past five years, Lief has highlighted repeatedly the consistent refusal of medical schools in this country to instruct in human sexual physiology, and, in doing so, he personally has been responsible for the most sweeping change in medical curriculum developed in the last two decades. Calderone has taken a pioneer step in sex education at both lay and professional levels with the development of the Sex Information and Education Council of the United States. Both of these physicians have given positive support to the growing demand that medicine accept its responsibility and educate its own, young and old alike. As in all things, there must be a beginning. There must be some way to teach the teachers.

Nor have the behavioral sciences failed to note and to reject science's sole timidity. Freud was well aware that his hypotheses lacked fundamental physiologic support when he wrote "It should be made quite clear that the uncertainty of our speculation has been greatly increased by the necessity of borrowing from the Science of Biology. Biology is truly a land of unlimited possibilities. We may expect it to give us the most surprising information and we cannot guess what answers it will return in a few dozen years to the questions we have put to it. They may be a kind which will blow away the whole of our artificial structure of hypothesis."

If problems in the complex field of human sexual behavior are to be attacked successfully, psychologic theory and sociologic concept must at times find support in physiologic fact. Without adequate support from basic sexual physiology, much of psychologic theory will remain theory and much of sociologic concept will remain concept.

There is every evidence from rapidly increasing individual and community-oriented pleas for aid, directed to all counseling resources, that our culturally induced sexual instability has gone far beyond the limited abilities of the concerned professions to cope. Nizer has written that the greatest single cause for family-unit destruction and divorce in this country is a fundamental sexual inadequacy within the marital unit. Kirkendall has stated that "if traditional morality no longer serves as a curb, neither does fear of consequence." How can biologists, behaviorists, theologians, and educators insist in good conscience upon the continued existence of a massive state of ignorance of human sexual response, to the detriment of the well-being of millions of individuals? Why must the basics of human sexual physiology create such high levels of personal discomfort among the very men and women who are responsible for guiding our culture? There is no man or woman who does not face in his or her lifetime the concerns of sexual tensions. Can that one facet of our lives, affecting more people in more ways than any other physiologic response other than those necessary to our very existence, be allowed to continue without benefit of objective, scientific analysis?

Why then must science and scientist continue to be governed by fear—fear of public opinion, fear of social consequence, fear of religious intolerance, fear of political pressure, and, above all, fear of bigotry and prejudice—as much within as without the professional world?

Van de Velde and Dickinson first dared to investigate and to write of sexual physiology. Yet they were forced to wait until the twilight of their professional careers before challenging public and professional opinion. Obviously, they were shocked when, aside from the expected opprobrium and implication of prurience, the biologic and behavioral sciences emphatically shut the door of investigative objectivity. Possibly history will record as Kinsey's greatest contribution the fact that his incredible effort actually enabled him to put his foot firmly in this door despite counterpressures that would have destroyed a lesser man.

This text represents the first step, a faltering step at best, but at least a first step toward an open-door policy. The door of investigative objectivity must not be closed again.

W. H. M.
V. E. J.

St. Louis

ACKNOWLEDGMENTS

The contributions and the contributors to this investigation and to this book are almost endless. Without their personal, professional, and monetary support, the work never would have achieved origin or substance.

There are those whose special skills we must acknowledge: William S. Sleator of the Department of Medical Physiology, K. Cramer Lewis and Marilyn Harris of the Department of Medical Illustration, and Sallie Schumacher of the Department of Psychology, all of Washington University, and Irene Gossage, our secretary.

To these friends and to all who have contributed, our respect and our gratitude.

W. H. M.
V. E. J.

CONTENTS

RESEARCH IN SEXUAL RESPONSE

I

THE SEXUAL RESPONSE CYCLE

In 1954 an investigation of the anatomy and physiology of human sexual response was initiated within the framework of the Department of Obstetrics and Gynecology of Washington University School of Medicine. A closely coordinated clinical-research program in problems of human sexual inadequacy was instituted in 1959. Since January, 1964, these programs have been continued under the auspices of the Reproductive Biology Research Foundation. During the past decade the anatomy of human response to sexual stimuli has been established, and such physiologic variables as intensity and duration of individual reaction patterns have been observed and recorded. Intensive interrogation (medical, social, psychosexual backgrounds) of both laboratory-study subject and clinical-research populations has been a concomitant of the basic science and clinical investigative programs since their inception. Material of significant behavioral content derived from these interviews will be presented in general rather than in statistical discussions.

Kinsey and co-workers published a monumental compilation of statistics reflecting patterns of sexual behavior in this country from 1938 to 1952. These reports of human sexual practices obtained by techniques of direct interrogation offer an invaluable baseline of sociologic information. Future evaluation of the work may reveal its greatest contribution to be that of opening the previously closed doors of our culture to definitive investigation of human sexual response.

Although the Kinsey work has become a landmark of sociologic investigation, it was not designed to interpret physiologic or psychologic response to sexual stimulation. These fundamentals of human sexual behavior cannot be established until two questions are answered: What physical reactions develop as the human male and female respond to effective sexual stimulation? Why do men and women behave as they do when responding to effective sexual stimulation? If human sexual inadequacy ever is to be treated successfully, the medical and behavioral professions must provide answers to these basic questions. The current study of human sexual response has been designed to create a foundation of basic scientific information from which definitive answers can be developed to these multifaceted problems.

The techniques of defining and describing the gross physical changes which develop during the human male's and female's sexual response cycles have been primarily those of direct observation and physical measurement. Since the integrity of human observation for specific detail varies significantly, regardless of the observer's training and considered objectivity, reliability of reporting has been supported by many of the accepted techniques of physiologic measurement and the frequent use of color cinematographic recording in all phases of the sexual response cycle.

A more concise picture of physiologic reaction to sexual stimuli may be presented by dividing the human male's and female's cycles of sexual response into four separate phases. Progressively, the four phases are: (1) the excitement phase; (2) the plateau phase; (3) the orgasmic phase; and (4) the resolution phase. This arbitrary four-part division of the sexual response cycle provides an effective framework for detailed description of physiologic variants in sexual reaction, some of which are frequently so transient in character as to appear in only one phase of the total orgasmic cycle.

Only one sexual response pattern has been diagrammed for the human male (Fig. 1-1). Admittedly, there are many identifiable variations in the male sexual reaction. However, since these variants are usually related to duration rather than intensity of response, multiple diagrams would be more repetitive than informative. Comparably, three different sexual response patterns have been diagrammed for the human female (Fig. 1-2). It should be emphasized that these patterns are simplifications of those most frequently observed and are only representative of the infinite variety in female sexual response. Here, intensity as well as duration of response are

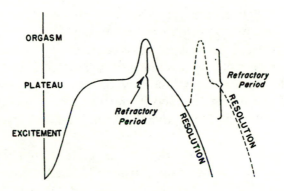

FIGURE 1-1

The male sexual response cycle.

factors that must be considered when evaluating sexual reaction in the human female.

The first or excitement phase of the human cycle of sexual response develops from any source of somatogenic or psychogenic stimulation. The stimulative factor is of major import in establishing sufficient increment of sexual tension to extend the cycle. If

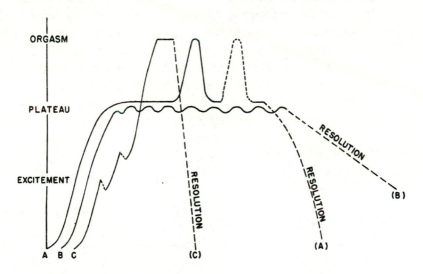

FIGURE 1-2

The female sexual response cycle.

the stimulation remains adequate to individual demand, the intensity of response usually increases rapidly. In this manner the excitement phase is accelerated or shortened. If the stimulative approach is physically or psychologically objectionable, or is interrupted, the excitement phase may be prolonged greatly or even aborted. This first segment and the final segment (resolution phase) consume most of the time expended in the complete cycle of human sexual response.

From excitement phase the human male or female enters the second or plateau phase of the sexual cycle, if effective sexual stimulation is continued. In this phase sexual tensions are intensified and subsequently reach the extreme level from which the individual ultimately may move to orgasm. The duration of the plateau phase is largely dependent upon the effectiveness of the stimuli employed, combined with the factor of individual drive for culmination of sex tension increment. If either the stimuli or the drive is inadequate or if all stimuli are withdrawn, the individual will not achieve orgasmic release and will drop slowly from plateau-phase tension levels into an excessively prolonged resolution phase.

The orgasmic phase is limited to those few seconds during which the vasoconcentration and myotonia developed from sexual stimuli are released. This involuntary climax is reached at any level that represents maximum sexual tension increment for the particular occasion. Subjective (sensual) awareness of orgasm is pelvic in focus, specifically concentrated in the clitoral body, vagina, and uterus of the female and in the penis, prostate, and seminal vesicles of the male. Total-body involvement in the response to sexual tensions, although physiologically well-defined, is experienced subjectively on the basis of individual reaction patterns. There is great variation in both the intensity and the duration of female orgasmic experience, while the male tends to follow standard patterns of ejaculatory reaction with less individual variation.

The human male and female resolve from the height of their orgasmic expressions into the last or resolution phase of the sexual cycle. This involutionary period of tension loss develops as a reverse reaction pattern that returns the individual through plateau and excitement levels to an unstimulated state. Women have the response potential of returning to another orgasmic experience from any point in the resolution phase if they submit to the reapplication of effective stimulation. This facility for multiple orgasmic expression is evident particularly if reversal is instituted at plateau tension level. For the man the resolution phase includes a superimposed refractory period which may extend during the involutionary phase as far as a lower excitement level of response. Effective restimulation to higher levels of sexual tension is possible only upon termination of this refractory period. With few exceptions, the physiologic ability of the male to respond to restimulation is much slower than that of the female.

Physiologic residuals of sexual tension usually are dissipated slowly in both the male and female unless an overwhelming orgasmic release has been experienced. Total involution is completed only after all manner of sexual stimuli have been withdrawn.

It always should be borne in mind that there is wide individual variation in the duration and intensity of every specific physiologic response to sexual stimulation. Those that occur early in the response cycle and continue without interruption during several phases are obvious (penile erection or vaginal lubrication). However, some physiologic reactions are fleeting in character and may be confined to one particular phase of the cycle. Examples are the plateau-phase color changes of the minor labia in the female and the coronal engorgement of the penis in the male.

In brief, the division of the human male's or female's cycle of sexual response into four specific phases admittedly is inadequate for evaluation of finite psychogenic aspects of elevated sexual tensions. However, the establishment of this purely arbitrary design provides anatomic structuring and assures inclusion and correct placement of specifics of physiologic response within the sequential continuum of human response to effective sexual stimulation.

The basic physiologic responses of the human body to sexual stimulation are twofold in character. The primary reaction to sexual stimuli is widespread vasocongestion, and the secondary response is a generalized increase in muscle tension. The vasocongestion may be either superficial or deep in distribution, and the myotonia reflected by voluntary or involuntary muscle contractions. The more severe vasocongestive and myotonic reactions are confined to plateau and orgasmic phases of the sexual cycle. There are multiple examples of these physiologic evidences of sex tension increment. They will be considered in discussions of specific body or organ system response, and, when presented individually, identified within the framework of reference provided by the four phases of the sexual response cycle.

Obviously, there are reactions to sexual tension that are confined by normal anatomic variance to a single sex. Also, basic differences develop between the two sexes in the intensity and duration of established reaction patterns. These differences will be underscored in context during the review of organ systems. However, again and again attention will be drawn to direct parallels in human sexual response that exist to a degree never previously appreciated. Attempts to answer the challenge inherent in the question, "What do men and women do in response to effective sexual stimulation?", have emphasized the *similarities, not the differences,* in the anatomy and physiology of human sexual response.

The question of why men and women respond as they do to effective sexual stimulation is not answered in this text. Neither the laboratory-study subject nor the clinical-research populations are sufficiently representative of the general population to allow definitive conclusions to be supported from behavioral material drawn from these groups and reported in the text.

Material of behavioral content drawn from the combined research populations will be presented as a contribution to the understanding of human sexual response patterns, but the prejudiced source of this information always must be borne in mind. Until a representative cross-section of the general population can be made available to research interests, even admittedly prejudiced information is of inordinate value in the study of human sexual behavior.

The widespread problems of human sexual inadequacy will not be attacked effectively by either medical or behavioral personnel until more definitive information is accumulated. Such data only will become available as the mores of our society come to accept objective research in human sexuality.

2

THE RESEARCH
POPULATION

A population of adult men and women who were willing to serve as subjects in the reproductive biology laboratory provided the opportunity for observations of anatomic and physiologic response to effective sexual stimulation made and recorded during this investigation. From these observations and from concomitant psychosexual interrogation, conclusions have been drawn regarding the sexual behavior of this selected population. Therefore, the general classification, history of development, and specific function of this group will be presented for evaluation.

Review of material presented by previous investigators will reveal an obvious prejudice in subject selection that unfortunately is inherent in any attempt to investigate biologic or behavioral facets of human sexual response. Cultural attitudes and residual sexual taboos always have inhibited statistically ideal population sampling. This study proves no exception to the general rule.

The investigation has underscored many examples of socially or culturally oriented situations that could influence human sexual response patterns. The possibility that the artificial atmosphere of the research laboratory might alter physiologic as well as psychologic sexual reaction patterns has been a constant concern. Natural anxiety of individual subjects for personal anonymity and concern with pressures of performance created problems that required special handling. The development of a range of interview techniques effective for all ages for both laboratory-study subject and clinical-research populations occupied hours of early research time. These are but a few of the multifaceted problems faced during the past eleven years.

In view of the investigative problems created by the technique of

9

working directly with sexually responding men and women, every attempt will be made to emphasize in context inadequacies of experimental control. When possible, countermeasures were taken to improve the sociologic, psychologic, and physiologic integrity of recorded data. They also will be explained in context.

First, the two questions to be answered: What happens to the human male and female as they respond to effective sexual stimulation? Why do men and women behave as they do when responding to effective sexual stimulation? In order to approach these problems, preliminary concentration was focused on the development of effective techniques of interrogation, observation, and physical recording. Once these technical tools were at hand, the next steps were the constitution of an adequate study-subject population and the integration of these volunteers into the investigative design.

The initial selection of individuals for study in the investigation of human sexual response was made from the prostitute population. This socially isolated group was regarded as knowledgeable, cooperative, and available for study. Availability was the determining factor during the initial stages of the program. It was presumed, at that time, that study subjects from more conservative segments of the general population would not be available (a presumption which later was proved to be entirely false).

For the first twenty months of the program, a total of 118 female and 27 male prostitutes contributed their sociosexual, occupational, and medical histories to the investigation. Ultimately, a small number of the total group (8 women and 3 men) were selected for anatomic and physiologic study. The criteria for selection were obvious intelligence, diverse experience in prostitution, ability to vocalize effectively, and, of course, a consistently high degree of availability and cooperation.

Suggestions by this select group of techniques for support and control of the human male and female in situations of direct sexual response proved invaluable. They described many methods for elevating or controlling sexual tensions and demonstrated innumerable variations in stimulative technique. Ultimately many of these techniques have been found to have direct application in therapy of male and female sexual inadequacy and have been integrated into the clinical research programs. This small group also served as laboratory-study subjects during the trial-and-error periods required initially to devise and to establish the investigative techniques subsequently employed throughout the study.

The interrogative material and experimental results derived from the prostitute population have not been included in the material being presented. Two factors influenced this decision: (1) The migratory tendencies of this population discouraged the recording of individual study-subject response patterns over extended periods of time, and (2) the varying degrees of pathology of the reproductive organs usually present in this population precluded the possibility of establishing a secure baseline of anatomic normalcy. Faced with the experimental necessity of developing a group of study subjects whose reproductive viscera could be related to baselines of anatomic normalcy and from whom long-range cooperation was possible, volunteers were sought from relatively selected social, intellectual, and economic backgrounds.

The study-subject population as finally constituted for this investigation has been established from selected segments of a metropolitan community. More specifically, it has been developed primarily from and sustained by the academic community associated with a large university-hospital complex. The concentration of study subjects from upper socioeconomic and intellectual strata provided by this major source of supply has not been offset by a statistically significant number of lower-range family units obtained from out-patient clinic sources.

There have been other sources of subject recruitment. A number of family units, initially presenting clinical problems either of sexual inadequacy or conceptive inadequacy, subsequently became a part of the study-subject population. They provide a wide range of sociogeographic and educational backgrounds. In recent years as knowledge of the work in progress spread locally, volunteers of all ages came from all social strata, and from a wide variety of educational backgrounds. During the past seven years recruitment has been limited primarily to specific project interests, such as the geriatric or the pregnancy-sexual response programs. Just over 75 percent of all male and female study subjects volunteered their services during this seven-year period. However, as has been obvious from the outset, the study-subject population represents many different forms of selectivity.

As discussed, the sample was weighted purposely toward higher than average intelligence levels and socioeconomic backgrounds. Further selectivity was established by an extensively detailed intake interview designed to determine willingness to participate, facility of sexual responsiveness, and ability to communicate finite details of sexual reaction. A physical examination sufficient to establish essential normalcy of the reproductive viscera also was required. All individuals with sociosexual aberrancy or grossly abnormal reproductive viscera were eliminated, further emphasizing the degree of selection inherent in the research population.

It would have been physically and financially impossible to maintain as a static force a study-subject population of the magnitude of the one presently reported. Many family units cooperated to the extent of recording their sexual response patterns, but were not retained thereafter as active members of the study-subject group. Other family units have remained available to the total program for years. Their reactions to effective sexual stimulation have been recorded with regularity to determine whether full familiarity with the program, the recording techniques, the investigative personnel, and the artificial environment of the research laboratory would in time alter their basic sexual response patterns.

During the years of continued cooperation, the ages, marital status, and even the parity of some female study subjects changed. The statistics as reported represent the status of the individual at the initial interrogation prior to activation in any of the research programs. A total of 382 women have cooperated actively with the investigation. The age range of female study subjects is from 18 to 78 years (Table 2-1). As might be expected, most of them (321 women, or 84 percent) were in the 18–40-year age bracket, with the largest concentration of active participants in the 21–30-year age group (182, or 47.6 percent). There have been two girls active in the program under 21 years of age—an 18-year-old who had been married three years, had one child, and worked with her husband as a family unit, and a 20-year-old who was recruited specifically for the artificial-vagina studies (see Chapter 7).

While to date the majority of the study-subject population has

TABLE 2-1

*Female Population of 382 Active Study Subjects ***

Age	No. Selected		No. Interviewed
18–20	2		2
21–30	182 ⎫		
31–40	137 ⎬ 346		460
41–50	27 ⎭		
51–60	23 ⎫		
61–70	8 ⎬ 34		157
71–80	3 ⎭		
Totals	382		619

* Data from prostitute population not included.

been female, 312 males have cooperated with the program on at least one occasion (Table 2-2). Their ages have ranged from 21 to 89 years. The majority of cooperative male study subjects (231, or 74 percent) were 21 to 40 years of age, with the largest number of active participants in the 21–30-age group (120, or 38.5 percent).

From an educational standpoint the research population has always been weighted toward a higher standard of formal training than would be true for a cross-section of the total population.

TABLE 2-2

*Male Population of 312 Active Study Subjects ***

Age	No. Selected		No. Interviewed
21–30	120 ⎫		
31–40	111 ⎬ 273		409
41–50	42 ⎭		
51–60	19 ⎫		
61–70	14 ⎪		
71–80	4 ⎬ 39		245
81–90	2 ⎭		
Totals	312		654

* Data from prostitute population not included.

Tables 2-3 and 2-4 list the educational backgrounds of the female and the male study subjects respectively. For purposes of statistical simplicity, listing in the *High School* column is dependent only upon matriculation. There were no active study subjects whose formal education did not include matriculation in high school. Dropouts are not listed separately. Both college and postgraduate training have been handled in similar fashion.

As would be expected, the level of formal education was consistently higher for male than female study subjects. Particularly was this level of education evident in study subjects past 50 years of age. Obviously, the inordinately high percentage of study subjects with exposure to postgraduate training is a direct reflection of both the metropolitan area and the hospital-university complex from which a dominant percentage of the total study-subject population has been drawn.

Although the research population purposely was weighted toward average or above-average intelligence, some of the study subjects were of less favorable backgrounds. Thirty-seven family units were drawn from a clinic population. The maximum of formal training in this group, four years of high school education, had been achieved by less than half these subjects.

There were 11 Negro family units, 3 of privileged and 8 of underprivileged backgrounds, included in the study-subject population.

TABLE 2-3

Education Among Female Study Subjects

Age	No.	High School *	College *	Postgraduate *
18–20	2	2	0	0
21–30	182	61	83	38
31–40	137	53	57	27
41–50	27	13	12	2
51–60	23	14	8	1
61–70	8	6	2	0
71–80	3	3	0	0
Totals	382	152	162	68

* Listing dependent only upon matriculation (highest level).

TABLE 2-4

Education Among Male Study Subjects

Age	No.	High School *	College *	Postgraduate *
21–30	120	25	49	46
31–40	111	19	41	51
41–50	42	5	17	20
51–60	19	1	7	11
61–70	14	1	9	4
71–80	4	0	3	1
81–90	2	0	2	0
Totals	312	51	128	133

* Listing dependent only upon matriculation (highest level).

In addition, two Negro women were evaluated without marital partners. One was a surgical castrate and the other had an artificial vagina. In view of the small number of Negro families in the study-subject population, it is obvious that the population has, over the years, been weighted toward the Caucasian rather than the Negro race.

From onset, no attempt has been made to maintain an accurate count of the number of male and female sexual response cycles experienced by study subjects in the environment of the research laboratory. However, a conservative estimate of 10,000 complete cycles of sexual response for the total research population certainly could be supported. Such an estimate would represent at least a 3:1 female dominance. Thus, a minimum of 7,500 complete cycles of sexual response have been experienced by female study subjects cooperating in various facets of the research program, as opposed to a minimum total of 2,500 male orgasmic (ejaculatory) experiences.

Both married and unmarried men and women have been included in the study-subject population of 382 women and 312 men. Over the years, 276 married couples have worked actively in the various programs. There have been 106 women and 36 men who were not married at the onset of their active cooperation with the program. Many had been married previously (81 women and 17 men) but were without a spouse immediately prior to joining the

TABLE 2-5

Reproductive Organ Pathology in Female Study Subjects

	Age-Group Distribution						
	18–20	21–30	31–40	41–50	51–60	61–70	71–80	Total
Pathology							
Cystoceles	0	1	5	3	2	2	0	13
Urethroceles	0	0	4	2	2	1	0	9
Rectoceles	0	2	4	3	2	2	0	13
Myomatous uteri	0	1	3	6	2	2	1	15
Retroverted uteri	0	17	13	6	6	3	0	45
Pelvic and labial varicosities	0	0	9	4	5	1	0	19
Pelvic endo-metriosis	0	3	2	1	0	0	0	6
Additional Data							
Vaginal agenesis (corrected)	1	5	1	0	0	0	0	7
Pregnancy							
Primaparas	0	2	1	0	0	0	0	3
Multiparas	0	1	2	0	0	0	0	3
Total active participants	2	182	137	27	23	8	3	382

study-subject group. Obviously, investigative programs oriented to a specific sex, such as work with the artificial vagina, pathologic clitoral hypertrophy, intravaginal contraceptive testing, or the ejaculatory mechanism and testicular elevation reactions, did not necessitate an active marital status for the study subjects involved.

It should be stated in context that early in the investigation, the nonmarried group also provided opportunity for comparison-control studies with established marital units. The unrehearsed physiologic and anatomic response patterns of the unmarried were recorded and contrasted to the mutually conditioned and frequently stylized sexual response patterns of the marital units. This technique for experimental control was abandoned as soon as it was

established unequivocally that there is no basic difference in the
anatomy and physiology of human sexual response regardless of
the marital status of responding units.

While pelvic pathology such as that characteristic of the prostitute population arbitrarily was screened out of the study-subject
population, normally occurring anatomic variants in reproductive
viscera purposely were included in an effort to avoid reactive selectivity within both female and male study subjects (Tables 2-5 and
2-6). Cystoceles, urethroceles, and rectoceles were present in parous
study subjects. Myomatous uteri, retroverted uteri, pregnancy-in-

TABLE 2-6

Reproductive Organ Pathology in Male Study Subjects

| | Age-Group Distribution | | | | | | | |
	21–30	31–40	41–50	51–60	61–70	71–80	81–90	Total
Pathology								
Benign prostatic hypertrophies	0	0	3	2	1	0	1	7
Unilateral testicular atrophies	0	1	1	0	0	0	0	2
Undescended testicle	1	0	0	0	0	0	0	1
Inguinal hernia (nonsymptomatic)	1	0	0	0	0	0	0	1
Inguinal hernia (symptomatic)	0	1	0	0	0	0	0	1
Varicocele (symptomatic)	0	1	0	0	0	0	0	1
Additional Data								
Uncircumcised penises	2	7	6	8	8	2	2	35
Total active participants	120	111	42	19	14	4	2	312

duced pelvic and labial varicosities, and pelvic endometriosis also were present in the female population. In the male groups there were benign prostatic hypertrophies, unilateral testicular atrophies, an undescended testicle, a small, nonsymptomatic and a large, symptomatic inguinal hernia and, finally, a large clinically symptomatic varicocele.

Nineteen women and 6 men had experienced major pelvic surgery before joining the study-subject population (Tables 2-7 and 2-8). Seven women served as active study subjects in the artificial-vagina group (see Chapter 7). Six women cooperated actively with the study of sexual response in a pregnant state, although 111 pregnant women (9 unmarried) responded to interrogation in depth with material of behavioral content (see Part 2 of Chapter 10). There were 35 uncircumcised male study subjects (see Part 2 of Chapter 12).

The study-subject population of unusual interest has been the geriatric group. There were 34 women (see Table 2-1) aged 51 to 78 years and 39 men (see Table 2-2) aged 51 to 89 years who co-

TABLE 2-7

Surgical History of Female Study Subjects

Major Pelvic Surgery	Age-Group Distribution							
	18–20	21–30	31–40	41–50	51–60	61–70	71–80	Total
Abdominal hysterectomy	0	0	1	3	3	1	0	8
Vaginal hysterectomy	0	0	0	0	1	1	0	2
Anterior and posterior colporrhaphy	0	0	0	0	1	0	0	1
Salpingo-oophorectomy	0	0	1	1	0	0	0	2
Oophorectomy	0	2	1	0	0	0	1	4
Salpingectomy	0	1	0	0	0	0	0	1
Inguinal herniorrhaphy	0	0	0	1	0	0	0	1

TABLE 2-8

Surgical History of Male Study Subjects

Major Pelvic Surgery	Age-Group Distribution							
	21–30	31–40	41–50	51–60	61–70	71–80	81–90	Total
Prostatectomy								
Transurethral	0	0	1	1	0	0	0	2
Perineal	0	0	0	1	0	0	0	1
Inguinal herniorrhaphy	1	1	0	0	1	0	0	3

operated actively with the research program. Five of the men were married to women in the 41- to 50-year age group. These additional couples were included arbitrarily in the geriatric population. Three of these female partners had been surgically castrated and the remaining two women were two and three years postmenopausal at the outset of experimental cooperation. The arbitrary decision to include these five couples in the geriatric group brought to 39 the number of aging married couples evaluated in depth for anatomic and physiologic patterns of sexual response. Even though the number of active geriatric participants is small and represents a high degree of selectivity, their contribution has been large, for their cooperation has extended over four years of concentrated investigation of geriatric sexual response.

There never has been an adequate number of study subjects available to the investigative programs. There are many reasons for the restricted size of the research population. As might be expected, the major reason for the statistically inadequate number of study subjects is the problem of insufficient research funding. Some programs were restricted in number of participants on the basis of anatomic scarcity; for instance, the artificial-vagina studies. Others were limited by psychosocial restraints. The geriatric research is an excellent example of this particular problem. Inevitably, the investigation's total orientation to human sexual response is a delimiting factor in itself.

No attempt will be made to provide statistical analyses of the sexual-behavior content elicited from detailed intake interviews. When compared to the Kinsey data collected 15 to 25 years ago, the material returned from a total of 654 male and 619 female interviews in a selected population is too meager statistically to be of significance. If the total sociologic import of the massive Kinsey contribution is to be realized, the work must be repeated in similar exhaustive detail and not suborned by attempts at inadequate statistical comparison. Therefore, behavioral content will appear only in the format of the chapters of general clinical discussion and the degree of selection of the subject matter again will be emphasized in context.

Of major import is the unique opportunity created by the research environment to observe, to record, and to evaluate the patterns of physiologic and psychologic response to effective sexual stimulation in a small, arbitrarily selected segment of male and female society to a degree never possible previously in medical or behavioral environment. Rather than material returned from interviews or questionnaires being the lone source from which to draw conclusions concerning human sexual behavior, the material in the chapters oriented to clinical reaction has been drawn from direct observation of sexual response—interviews in depth of behavioral content, discussions of the individual's sexual response patterns, laboratory recording and analysis of reactive patterns, and so on. Clinical chapters such as those dealing with geriatric sexuality, sexuality and pregnancy, and sexuality of male and female study-subject populations will be compilations of interrogative material of behavioral content. In addition, observation and physiologic recording of their sexual response patterns will be reported.

Finally, and possibly most important, is the information gleaned from eleven years of opportunity to work directly with the human male and female responding to effective sexual stimulation. It constantly should be borne in mind that the primary research interest has been concentrated quite literally upon what men and women do in response to effective sexual stimulation, and why they do it, rather than on what people say they do or even think their sexual reactions and experiences might be.

Further, modes or means of sexual stimulation will be described without reservation at this point and not constantly referred to in the body of the text. Recorded and observed sexual activity of study subjects has included, at various times, manual and mechanical manipulation, natural coition with the female partner in supine, superior, or knee-chest position and, for many female study subjects, artificial coition in supine and knee-chest positions. No study subject has been able to fantasy to orgasm under observation.

Hundreds of complete cycles of sexual response have been accomplished under artificial coition with female study subjects in supine or knee-chest positioning. This research technique was created and has been used for three purposes. First, it provides opportunity for observation and recording of intravaginal physiologic response to sexual stimuli [215]. Second, the technique is employed to establish the effectiveness of intravaginal, mechanical, or chemical contraceptives without months or years of field trial wherein reported results of necessity are based upon the tragedy of pregnancy failure [129, 130, 134, 135]. Third, artificial coital techniques are used clinically to replace surgery in the creation of artificial vaginas in women born with vaginal agenesis (see Chapter 7). This use of the technique was suggested by the work of Frank and Geist [77, 78].

The artificial coital equipment was created by radiophysicists. The penises are plastic and were developed with the same optics as plate glass. Cold-light illumination allows observation and recording without distortion. The equipment can be adjusted for physical variations in size, weight, and vaginal development. The rate and depth of penile thrust is initiated and controlled completely by the responding individual. As tension elevates, rapidity and depth of thrust are increased voluntarily, paralleling subjective demand. The equipment is powered electrically.

Orientation to this equipment obviously was necessary for study subjects with established coital and/or automanipulative experience. The orientation periods provided opportunities for evaluating subjective fantasy and conditioning processes employed by study subjects for sex tension increment.

In view of the artificial nature of the equipment, legitimate issue may be raised with the integrity of observed reaction patterns. Suffice it to say that intravaginal physiologic response corresponds in every way with previously established reaction patterns observed and recorded during hundreds of cycles in response to automanipulation.

Homosexual material, although recorded in both behavioral and physiologic context for both sexes, has not been included in this text. The returns from this facet of human sexual response are too inadequate at present to warrant consideration. At the present pace investigative maturity will not be reached in this program for at least another four to five years.

Once selected as study subjects, males and females were exposed to a controlled orientation program before assuming active participation as members of the research population. Detailed medical, social, and sexual histories were obtained from each study subject by both male and female interrogators. The technique of dual-sexed interrogation was designed to satisfy two purposes. First was the necessity to orient a prospective study subject to the fact of bisexual supervision of all investigative procedures, and second was the demand for security of factual reporting. Material of sexual connotation has been elicited from study subjects more effectively and accurately by interview teams with both sexes represented than by single-sexed interrogation [132, 133, 210, 216]. The exposure to the prostitute population emphasized the advantages of this technique. History-taking also served secondarily as a step in preparation for active participation in the program. It acquainted the study subject with the investigative personnel and established in his or her mind the investigators' nonjudgmental attitudes and their authoritative roles.

The next step in the orientation program was to demand that potential study subjects undergo physical examinations to rule out any gross pathology, particularly of the reproductive organs.

The individuals considering active cooperation with the program then were exposed to the research quarters. All equipment was exhibited and its function explained to the uninitiated. Sexual activity first was encouraged in privacy in the research quarters and then continued with the investigative team present, until the study subjects were quite at ease in their artificial surroundings. No attempt was made to record reactions or introduce other members of the research personnel to the reacting unit, until the study subjects felt secure in their surroundings and confident of their ability to perform. They rapidly gained confidence in their ability to respond successfully while subjected to a variety of recording techniques. Finally, this period of training established a sense of security in the integrity of the research interest and in the absolute anonymity embodied in the program.

Once total confidence was attained, the study subjects were directed to the particular phase of the overall program in which their cooperation was considered to be of greatest value. These areas have included the specific and long-continued recording of anatomic and physiologic response to effective sexual stimulation, pregnancy sexuality, geriatric sexuality, etc. Frequently, one recording session for a family unit was sufficient to demonstrate conclusively the anatomic variations and physiologic reactions that will be described in this text. In other instances family units have been immersed in multiple facets of the investigative program and have cooperated actively for many years.

FEMALE SEXUAL RESPONSE

3

FEMALE EXTRAGENITAL
RESPONSE

The human female's physiologic response to sexual stimulation is not confined to the reproductive viscera. Sexual tensions involve many areas other than the primary or secondary organs of reproduction. Although previous investigators [56, 67, 76, 143, 144] have described extragenital reactions, the extent of influence of increasing or unresolved sexual tensions upon the corporate body structure has not been appreciated. A description of specific organ or system response may serve to highlight this influence of sexual tensions on the total body economy.

Physiologic reactions to sexual stimulation are superficial and/or deep vasocongestion and both generalized and specific myotonia. The vasocongestive reactions relate to any phase of the sexual cycle, while muscle tensions usually become clinically obvious during the plateau phase. An example of superficial vasocongestion is the sex flush appearing on body surfaces, while deep vasocongestion is demonstrated by the development of the orgasmic platform. Target organs such as the breasts provide evidence of a combination of both superficial and deep vasocongestion. Muscle groupings which reflect a generalized tension response to sexual stimulation are those of the hands, feet, and abdomen, while the bulbospongiosus and ischiocavernosus muscles and the rectal sphincter provide examples of specific muscle tension.

This chapter will be devoted to evidence of physiologic response to effective sexual stimulation in other than the target organs of female reproduction. The four phases of the cycle of sexual response will be employed as a descriptive aid.

27

THE BREASTS

EXCITEMENT PHASE

Nipple erection is the first evidence of the breasts' response to sex tension increment (Fig. 3-1). This erective reaction occurs as the result of involuntary contraction of muscular fibers within the structure of the nipple [94]. Frequently the nipples do not achieve full erection simultaneously. One nipple may become fully erect and tumescent, while the other lags in erective rapidity and tumescent size. Inverted nipples may reverse their quiescent station to assume a position of semierection, or if the inversion is irreducible there may be no nipple indication of the breast's reaction to sexual stimulation.

Full erective response may increase nipple length by 0.5–1.0 cm. and base diameter by 0.25–0.5 cm. over unstimulated measurements. Large, protruding nipples usually have relatively less capacity for size increase than do the more normal-sized nipples. Excessively small nipples have little physiologic capacity to respond to sexual stimulation with a measurable increase in size.

A second physiologic alteration developing during the excitement

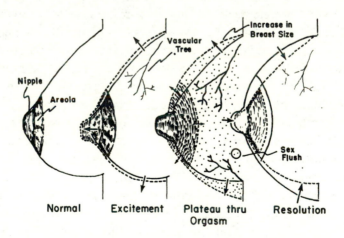

FIGURE 3-1

The breasts in the female sexual response cycle.

phase is that of increased definition and extension of the venous patterns of the breast. If the breast is of sufficient size, venous engorgement of the inferior surfaces occurs, but may not be well defined until somewhat later in the cycle. The larger breasts usually display the most definitive expansion of venous pattern. Engorgement of the vascular tree of the breast rarely extends centrally as far as the areola, since there is fairly rapid venous drainage to the axillary and internal mammary veins [54, 299].

As sexual tensions progress toward plateau-phase levels, there is an obvious increase in the actual size of the breasts. This size increment results from the organs' deep vasocongestive reaction. When the responding woman is in an erect position, the engorgement is easily visualized in the lower or inferior portion of a pendulous breast. When the responding woman is supine, the overall increase in breast size is more apparent. Tumescence of the female breast with sexual excitement was first described by Dickinson [56] over thirty years ago.

PLATEAU PHASE

Marked areolar engorgement develops late in the excitement phase (see Fig. 3-1). The areolae become so tumescent with plateau-phase tensions that they impinge upon the erect nipples, creating an illusion that the responding woman partially has lost nipple erection. This impression is corrected as resolution-phase areolar detumescence again reveals the erect nipples.

Before the woman experiences the final physiologic surge toward orgasm, the unsuckled breast will have increased in size by one-fifth to one-fourth over unstimulated, baseline measurements. The breast that has been suckled usually does not demonstrate as definitive a size increase as the unsuckled breast. This anatomic variant may be the result of the marked hypertrophy of venous drainage which milk production develops in the suckled breast. Suckling increases venous drainage and tends to minimize the deep vasocongestive effects of sexual tension.

When more than one child has been suckled, the breasts rarely demonstrate significant increase in size under the influence of sexual stimulation. Apparently the increase in breast size stimulated by sexual tensions not only is related to the physiologic response of vasocongestion but also may be implemented by the fibrous-tissue elements which invest breast lobules as supportive aids [94]. Overdistention of the breast so frequently seen in the early phases of milk production tends to impair the effectiveness of these supportive fibrous-tissue elements. It is understandable, therefore, that women who have suckled children frequently show little or no increase in breast size under sex tension influence.

As the woman elevates from excitement-phase to plateau-phase levels of sexual tension, a pink mottling frequently appears over the anterior, lateral, and, ultimately, the inferior surfaces of the breasts. Actually, this maculopapular rash first appears over the epigastrium and spreads to the breast surfaces as the plateau phase is firmly established. This superficial vasocongestive reaction of the breast will be considered in detail in the discussion of the sex-flush phenomenon.

ORGASMIC PHASE

There is no specific breast reaction to the experience of orgasm. Nipple erection and areolar tumescence have been established, the vascular tree stands out in bold relief, the nonsuckled breast is increased significantly in size over its unstimulated baseline, and the sex flush is well established. Maturity of superficial and deep vasocongestive responses is concentrated in a reactive peak during the orgasmic experience.

RESOLUTION PHASE

Advent of the resolution phase is signalled by the rapid disappearance of the sex flush and the simultaneous detumescence of the areolae (see Fig. 3-1). With areolar detumescence the nipples appear to regain full erection. The impression is created that they are undergoing a secondary erective reaction, either from new stimulative influence or retained stimulative effect. This reaction has been termed a "false erection" and is simply the result of early detumescence of the grossly engorged areolae which occurs well before nipple erection subsides.

As a general rule, the nonsuckled breasts lose their deep vaso-congestion slowly. Frequently, the increment in breast volume is retained for five to ten minutes after the orgasmic phase has been terminated, while superficial venous patterns on the breast surfaces may persist even longer. Erect nipples may undergo complete involution before the venous patterns return to their normally ill-defined state. This persistence of both superficial and deep vasocongestive influences, particularly evident in the non-suckled breast, may be the result of plateau-phase overdistention of the plexus venosus areolaris (circulus of Haller) [94]. This over-distention results in slowed venous drainage into the internal mammary veins during the resolution phase. (For discussion of the pregnant and lactating breast, see Part 1 of Chapter 10.)

THE SEX FLUSH

The protean character of the sex-flush reaction to effective sexual stimulation has not been considered previously. Its generalized distribution was emphasized when illumination necessary for successful laboratory cinematography increased the skin temperature and more clearly defined the sex flush. Both intensity and distribution patterns of the sex flush vary among individuals, but as a rule the severity of the flush reaction may be considered a direct indication of the intensity of the sexual tensions experienced by the responding woman.

This maculopapular type of erythematous rash first appears over the epigastrium either late in the excitement or early in the plateau phase of the sexual cycle. The sex flush then spreads rapidly over the breasts, first appearing on the anterior and superior surfaces of the breasts and then on the anterior chest wall. The lateral and then the medial breast surfaces become involved. Finally, as orgasm is imminent, an extension of the flush is often noted on the undersurfaces of the breasts.

On occasion most of the body surfaces of sexually responding women have given evidence of this superficial vasocongestive reaction during the plateau phase of the sexual cycle. The sex flush may be observed spreading over the lower abdomen, the shoulders, and even the antecubital fossae as sexual tensions mount. With impending orgasm, this measles-like rash even may spread over the anterior and lateral borders of the thighs and over the buttocks and the entire back.

The sex flush reaches a peak of color concentration and its widest distribution late in the plateau phase and terminates abruptly with orgasmic experience. The flushed, strained features of the human female as she reaches for orgasmic release of her overwhelming sexual tensions have been described graphically in the past [30, 88, 141, 144, 305]. It now is evident that the vascular flush previously described as confined to the face and anterior chest wall frequently may have widespread body distribution.

The sex flush extends over the body surfaces of susceptible individuals, paralleling in intensity the severity of impending orgasmic experience. Approximately 75 percent of all women evaluated demonstrated the sex flush on occasion. It should be recalled that the women recorded in the study had well-established sexual response patterns with significant prior orgasmic experience. The figure of 75 percent for sex-flush appearance among female study subjects represents, in all probability, a higher incidence than that to be observed in the general population.

The sex flush disappears from the different body sites in almost opposite order from its sequential appearance. The maculopapular rash disappears quickly from the back, buttocks, lower abdomen, arms, and thighs. It is much slower to resolve from the chest, breasts, neck, face, and finally from its initial appearance site over the epigastrium.

MYOTONIA

Myotonia, which becomes clinically obvious late in the excitement phase and during plateau-phase levels of sexual tension, is both generalized and specific in character. Usually muscles contract with regularity or in spasm in an involuntary manner, but contraction frequently may be voluntary depending upon coital positioning. For example, carpopedal spasm [119, 144, 271, 319], a spastic contraction of the striated musculature of the hands and feet, frequently is present late in plateau or during orgasm for the female in supine coital positioning or during automanipulation. Carpopedal spasm is reflected involuntarily by fingers and toes that are not devoted voluntarily to clutching responses. (A detailed discussion of both male and female myotonia during sexual response has been undertaken in Chapter 18.)

URETHRA AND URINARY BLADDER

Recurrent observation of the urethra has shown occasional involuntary distention of the external meatus during orgasm. This dilatation of the urethral meatus is of minimal degree and occurs with no established regularity. The meatus returns to its normal constricted state before contractions of the vaginal orgasmic platform have ceased.

The urge to void during or immediately after intercourse has been reported previously [69, 93, 102, 319]. In most instances study subjects who complained of postcoital dysuria had a high, firm perineum and nulliparous constriction of the vaginal outlet. These nulliparous structures combine to hold and direct the penis along the anterior vaginal wall during mounting and active coition. Thus the posterior wall of the urinary bladder reflexly is irritated subsequent to repeated penile thrusting. Only with this particular type of outlet structuring was bladder or urethral tenesmus described as a coital or postcoital complication. The clinical symptoms resulting from this situation have been known to the medical profession for years and given the slang term of "honeymoon cystitis."

Three other study subjects, two with clinically symptomatic urethroceles and one with a symptomatic cystocele and second-degree uterine descensus, also have complained on occasion of postcoital dysuria. Since all three women have a constant level of residual urine in the bladder, the urge to void, particularly after either long-continued or severe coital activity, is readily understood.

THE RECTUM

Voluntary contraction of the external rectal sphincter together with the gluteal musculature may be employed during both excitement and plateau phases of sexual response. Many women use this stimulative technique when driving for sex tension increment.

Involuntary contraction of the rectal sphincter occurs only during an orgasmic experience. Although this reaction does not occur with consistency, it is a significant indication of the intensity of orgasm. During a more intense orgasmic response, the external rectal sphincter may contract two to five times. The contractions develop at 0.8-second intervals, as do those of the orgasmic platform in the outer third of the vagina (see Part 1 of Chapter 6). In short, the contractions of orgasm described for the orgasmic platform frequently are paralleled by simultaneous contractions of the perineal body and the external rectal sphincter. External rectal-sphincter contractions occur most frequently during an orgasm elicited by automanipulation but occasionally have been observed during coition..

HYPERVENTILATION

Hyperventilation develops late in the plateau phase of the female's sexual cycle, lasts through the entire orgasmic experience, and terminates early in the resolution phase. Respiratory rates over 40 per minute have been recorded at the apex of severe orgasmic experience. With a minimal-intensity orgasm lasting 3 to 5 seconds, clinically obvious hyperventilation may not occur.

TACHYCARDIA

The heart rate usually is elevated significantly during late plateau and orgasmic phases of the sexual cycle. Rates from 110 to 180+ beats per minute have been recorded. Higher heart rates reflect more variation in orgasmic intensity for the female than for the

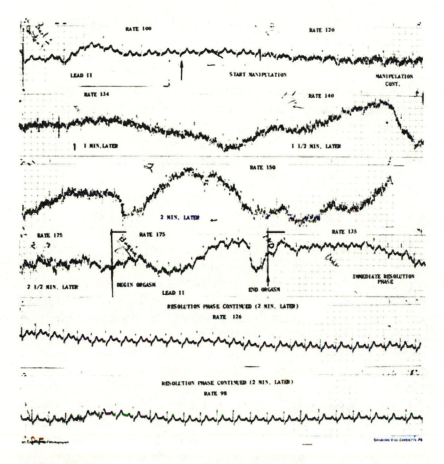

FIGURE 3-2

Electrocardiogram (Lead II) of female study subject. Note cardiac rate of 175/min. late in plateau and during orgasm.

male. The highest cardiac rates have been returned during female masturbatory sequences rather than during coition (Fig. 3-2).

BLOOD PRESSURE

Elevations of systolic pressure of 30 to 80 mm. Hg have been recorded from female study subjects late in plateau or during

orgasmic experience. Diastolic pressure elevations are usually in
the range of 20 to 40 mm. Hg. With minimal-intensity orgasmic
experience, diastolic pressure readings essentially may be unaffected.

There have been several attempts to evaluate cardiorespiratory
response to sex tension increasement in the past [22, 67, 151, 152,
220, 282]. Specifics of female cardiorespiratory response to elevated
sexual tensions will be published in a separate consideration of
both general body and target-organ physiology.

PERSPIRATORY   REACTION

Many women have described a sensation of being excessively
warm or feeling cold as they resolve their sexual tensions. Frequently, resolution of the sex flush occurs coincidentally with the
appearance of a widespread film of perspiration. An involuntary
perspiratory reaction may develop during the resolution phase regardless of the degree of physical activity demanded of the responding woman. This reaction has been described frequently
during the past thirty years [67, 239, 278, 319].

A filmy sheen of perspiration may appear over the back, thighs,
and anterior chest wall as the immediate postorgasmic woman
recovers conscious interest in her environment. Perspiration has
been noted to run from the axillae and also become an evident
surface coating of the entire body from shoulders to thighs. The
soles of the feet and palms of the hands are areas of perspiratory
concentration for some women. Perspiration appears on the fore-
heads and upper lips of women whose faces have been mottled
and swollen in the usual patchy manner immediately prior to
orgasmic tension release.

Approximately one-third of female study subjects display a tendency toward the perspiratory reaction. It appears simultaneously
over all body sites in the immediately postorgasmic time sequence
and is the first indication of resolution of the superficial vasocongestive response of the skin (sex flush) to effective sexual
stimulation. For instance, the breasts may be covered with perspiration, and still demonstrate disappearing vestiges of the sex-flush

phenomenon. The degree of obvious perspiratory coating of the body surfaces depends on opportunity for evaporation and/or absorption by clothing or bedding materials. The severity of the reaction when it occurs also parallels directly the intensity of the orgasmic expression.

If orgasmic-phase levels of sexual tension are not experienced, a generalized perspiratory reaction rarely occurs during the resolution phase. If a sex flush develops during the woman's plateau phase and she does not achieve orgasmic-phase release, the flush fades rapidly (once sexual stimulation is discontinued), usually resolving without evidence of perspiratory response other than on the palms of the hands or soles of the feet.

It constantly must be borne in mind that physiologic response to sex tension increment is a protean reaction. In addition to the selected responses highlighted in this chapter, generalized vasocongestion and myotonia involve many body areas and organ systems other than the target organs. Future investigation will establish specifics of response in such areas as the organs of special sense, the endocrine system, and the hypothalamic and higher cortical centers.

4

FEMALE EXTERNAL
GENITALIA

ANATOMY AND PHYSIOLOGY

The human female's external genitalia include the labia majora, the labia minora, and the clitoris. The anatomy, physiology, and psychodynamics of clitoral response will be presented in Chapter 5. A consideration of the function of Bartholin's glands, located in the labia minora, has been included in this discussion. The glands are intimately associated anatomically with the external genitalia, since they are imbedded in the labia minora. Both the external genitalia and Bartholin's glands respond to sexual stimulation through an interdependency on mutual neuroactivation (Fig. 4-1).

LABIA MAJORA

When unstimulated sexually, the major labia normally meet in the midline and provide positive protection for the subjacent structures—the minor labia, the vaginal outlet, and the urinary meatus. Obstetric trauma may cause loss in the integrity of major-labial midline accommodation and a subsequent loss of protection to the vaginal outlet. Thus, the response patterns of the major labia to effective sexual stimulation may be influenced markedly by the patient's obstetrical history. For this reason the labial response patterns will be discussed for both nulliparous and multiparous women. As sexual tensions rise for the nullipara during the excitement phase, the major labia thin out and flatten against the perineum. There is also a slight elevation of the labia in an

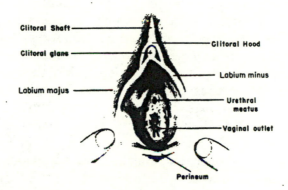

Clitoral Shaft
Clitoral glans
Labium majus
Clitoral Hood
Labium minus
Urethral meatus
Vaginal outlet
Perineum

FIGURE 4-1

The human female external genitalia.

upward and outward direction away from the vaginal outlet. This labial flattening and anterolateral elevation or displacement usually are completed late in the excitement phase or soon after plateau-phase levels of sexual tension are achieved.

The anatomic displacement of the major labia may be caused by protrusion of the rapidly engorging minor labia and preliminary vasocongestion of the external third of the vagina. There is an alternative possibility. The flattening and anterolateral elevation of the major labia away from the vaginal outlet may be an involuntary neurophysiologic attempt to remove any exterior impediment to the anticipated mounting process. There are no changes in the anatomy of the labia majora either late in plateau or during the orgasmic phase of the sexual cycle.

During the resolution phase the major labia return to their normal thickness and midline positioning. Labial involution occurs rapidly if the woman undergoes an orgasmic experience. If only plateau-phase levels of sexual tension are attained, an extended resolution phase usually is experienced, and the separated and engorged major labia may be slow to return to midline positioning. Additionally, if the nulliparous woman is long maintained in excitement or plateau phase, the separated labia majora become severely engorged with venous blood and sometimes even develop edema. Frequently this severe labial vasocongestion may persist for several hours after cessation of all sexual stimulation.

For the multiparous woman (particularly one with labial varicosities), the labia majora react to elevated sexual tensions in a somewhat different manner. Instead of flattening and undergoing excitement-phase anterolateral elevation against the perineum, the labia majora become markedly distended with venous blood. Occasionally, during an extended plateau phase, a two- to threefold increase in labial diameter has been demonstrated. If the vasocongestive increase in diameter occurs, there is no flattening or elevation of the swollen labia against the perineum; they hang pendulous and swollen as a partial curtain to the vaginal outlet. There is, however, a slight lateral movement away from the midline, so that even the swollen vasocongested major labia of a multiparous woman do not interfere with the normal mounting process. In fact, a restricted but discernible mounting invitation still occurs as an involuntary response to effective sexual stimulation regardless of the degree of multiparity of the individual.

In a varicosity-distressed woman, labial vasocongestion may persist through a two- to three-hour resolution phase before complete detumescence, presuming only plateau-phase levels of sexual tension have been attained. If an orgasm is experienced by these women, detumescent involution of labial vasocongestion is much more rapid. Understandably, the more advanced the varicosity involvement of the major labia, the more severe the vasocongestive reaction of the labia under sex tension influence.

LABIA MINORA

Aside from changes in the clitoris, the most definitive changes that develop in the external genitalia during a complete cycle of sexual response appear in the labia minora. With a well-established excitement phase the labia minora of both nulliparous and multiparous women expand markedly in diameter. When sexual tensions reach plateau-phase levels, the labia minora increase at least two, occasionally three, times in diameter. With expansion in diameter, the minor labia protrude through the protective curtain of the thinning major labia, and in so doing may be partially responsible for major labial separation and anterolateral elevation. The increase in minor labial diameter adds at least 1 cm. to the clinical length of the vaginal barrel during coition with the exception of the posterior wall of the vaginal outlet (the fourchette).

Once the excitement-phase vasocongestive increase in diameter has been completed, the minor labia are prepared for one of the most unique but specific physiologic reactions occurring in the human female during the cycle of sexual response. Vivid color changes develop in the engorged minor labia during the plateau phase of the sexual response cycle. These color changes may be equated with the parity of the individual. When the nulliparous woman reaches a plateau phase of sexual tension, the minor labia undergo a color change which ranges from a pink to a bright red color. This florid coloration diffuses along both sides of the vaginal outlet, usually including the clitoral hood in its progression. The multiparous individual also evidences an obvious color change which varies from the bright red to a deep wine color. As a rule the darker the color change in the minor labia, the more severe the degree of pelvic and labial varicosity involvement.

So specific are these plateau-phase color changes that the minor labia have been termed the "sex skin" of the sexually responding woman. No premenopausal woman has been observed to reach plateau-phase levels of sexual tension, develop the "sex skin" color changes, and then not experience an orgasm. In order to support this statement it must be presumed that the particular form of effective sexual stimulation which produced the sex-skin color changes is continued without major alteration. (The term *sex skin* should not be confused with terminology referring to circumperineal changes in female monkeys and apes during estrus [11].)

Many women have progressed well into plateau-phase levels of sexual response, had the effective stimulative techniques withdrawn, and been unable to achieve orgasmic-phase tension release. These women, if sufficiently excited, may have developed the sexskin color changes. When an obviously effective means of sexual stimulation is withdrawn and orgasmic-phase release is not achieved, the minor-labial coloration will fade rapidly, long before the vasocongestive increase in diameter is resolved.

It is obvious that the sex skin (labia minora) provides satisfactory clinical evidence of the degree of sexual tension experienced by the individual. When the sex-skin reaction develops, the woman has reached plateau-phase levels of physiologic response to her sexual tensions. Generally, the more brilliant and definitive the color change, the more intense the individual's response to the particular means of sexual stimulation. It should be emphasized that the development of the sex-skin reaction is clinically pathognomonic of impending orgasmic-phase expression. No woman ever has been observed to attain orgasmic release of sexual tensions without first demonstrating the specific minor-labial color changes.

During the resolution phase, the tension-induced coloration of the sex skin quickly devolves from deep or bright red to light pink, generally within 10 to 15 seconds after orgasmic expression. The second stage of color loss of the minor labia (pink to unstimulated colorless state) is also relatively rapid but does not occur with uniform regularity. The sex skin undergoing final or secondary loss of sex tension color has a rather blotchy appearance.

When the minor labia have achieved a plateau-phase color change, at least a twofold increase in diameter has developed. Together with the vasocongested outer third of the vagina (orgasmic platform) the minor labia form an engorged distal vaginal barrel and provide supportive containment for the penile shaft. In essence, the changes in the labia (majora and minora) that plateau-phase levels of sexual tension develop result in opening the vaginal outlet by removing the natural anatomic protection from the vaginal orifice. In addition, the sex-skin coloration (minor labia) signifies intense female sexual tensions and is clinically pathognomonic of impending orgasm, if the effective stimulative techniques are maintained.

BARTHOLIN'S GLANDS

Bartholin's glands are vulvovaginal glands located in each of the minor labia. They have ductal outlets on the inner surfaces of the labia, immediately adjacent to the vaginal introitus. The glands secrete a mucoid material that for many years has been presumed to contribute sufficient lubrication for successful and painless vaginal penetration [30, 119, 228, 268, 305, 318, 319].

It is true that Bartholin's glands do respond to sexual stimulation by secretory activity (Fig. 4-2). However, this secretory activity develops only late in excitement-phase or early in plateau-phase levels of sexual tension. The nulliparous study subjects rarely produce more than a drop of the mucoid material from each duct.

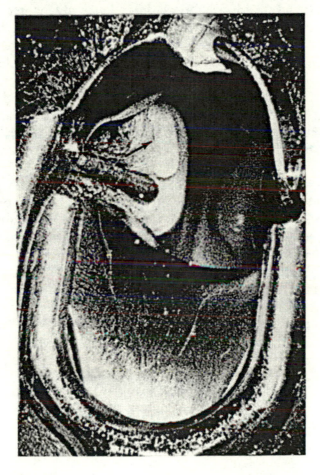

FIGURE 4-2
Specific secretory activity of Bartholin's gland at the plateau phase.

The multiparous woman occasionally develops 2 or even 3 drops of material. Under observation, however, there never has been sufficient secretory material produced to accomplish more than minimal lubrication of the vaginal introitus.

Basic vaginal lubrication develops in a transudation-like reaction through the walls of the vagina. This lubricating material appears early in the excitement phase, a matter of seconds after the onset of any form of sexual stimulation. This material normally is produced in such copious quantity that the vaginal barrel and outlet are quickly and effectively lubricated (see Part 1 of Chapter 6).

Bartholin's gland activity is stimulated most effectively by long-continued coital connection, particularly if the female partner is maintained in late excitement-phase or plateau-phase levels of sexual tension. In this situation, male coital stroking continued for lengthy periods of time stimulates the Bartholin's glands to contribute significantly to introital lubrication, but they never produce sufficient material to effect lubrication of the vaginal barrel. During automanipulative episodes there frequently is no evidence of Bartholin's gland secretory activity.

Bartholin's gland secretions have been assigned the role of reducing vaginal acidity to promote greater sperm longevity during intravaginal containment [119, 325]. However, the material secreted is so minute in amount when compared to that produced by the mechanism of vaginal lubrication that the concept is mechanically and chemically impossible. It is true that there is a recordable elevation of vaginal pH during episodes of long-continued sexual stimulation, but this change in vaginal acidity is the result of the production of vaginal lubrication and not Bartholin's gland secretory material (see Part 3 of Chapter 6).

In summary, increased secretory activity of Bartholin's glands is a negligible factor in vaginal-barrel or even introital lubrication. Not only is a minimal amount of material produced, but also the plateau-phase timing of such secretory activity rules out a role as the primary mechanism of vaginal lubrication.

5

THE CLITORIS

1. ANATOMY AND PHYSIOLOGY

The clitoris is a unique organ in the total of human anatomy. Its express purpose is to serve both as a receptor and transformer of sensual stimuli. Thus, the human female has an organ system which is totally limited in physiologic function to initiating or elevating levels of sexual tension. No such organ exists within the anatomic structure of the human male.

Conceptualization of the role of the clitoris in female sexual response has created a literature that is a potpourri of behavioral concept unsupported by biologic fact. Decades of "phallic fallacies" have done more to deter than to stimulate research interest in clitoral response to sexual stimulation. Unfortunately, the specific roles previously assigned clitoral function in female sexual response were designated by objective male consideration uninfluenced by and even uninformed by female subjective expression.

In the past, anatomic dissection, microscopic examination, and surgical ablation of the clitoris have established the organ as a homologue of the male penis [94, 262, 278]. The clitoris (Fig. 5-1) consists of two corpora cavernosa enclosed in a dense membrane primarily of fibrous-tissue origin. This capsule has recently been shown to contain elastic fibers and smooth-muscle bundles [49]. The fibrous capsules unite along their medial surfaces to form a pectiniform septum which is well interspersed with elastic and smooth-muscle fibers. Each corpus is connected to the rami of the pubis and ischium by a crus. The clitoris is provided (as is the penis) with a suspensory ligament which is inserted along the anterior surface of the midline septum. In addition, two small muscles, the ischiocavernosus muscles, insert into the crura of the clitoris and have origin bilaterally from the ischial rami.

The dorsal nerve of the clitoris is very small and is the deepest

45

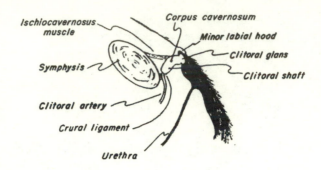

FIGURE 5-1

The clitoris in retraction (lateral view).

division of the pudendal nerve. It terminates in a plexus of nerve endings within the substance of the glans and the corpora cavernosa. Dahl described both myelinated and unmyelinated fibers of the somatic and vegetative nervous systems [48]. Pacinian corpuscles are distributed irregularly throughout the autonomic system nerve fibers both in the glans and the corpora but usually have greatest concentration in the glans [157, 158].

The pacinian corpuscles are concerned primarily with proprioceptive stimulation. The blood supply to the clitoris is derived from the deep and dorsal clitoral arteries, which in turn are branches of the internal pudendal artery. The arterial supply to and venous return from the clitoris follow the distribution patterns described for the penis (see Part 1 of Chapter 12). Although the blood supply has distribution patterns similar to that of the penis, clitoral vascularity obviously is accomplished from vessels of smaller capacity for fluid volume.

Anatomic dissection was supplemented by reported clinical mensuration of the female phallus only forty years ago. Clitoral glans size has been established at an average of 4 to 5 mm. in both the transverse and the longitudinal (less accurate) axis. One hundred adult females were used in Dickinson and Pierson's [59] first sample. Dickinson later described clitoral position with relation to the distance between the crural origins on the anterior border of the symphysis and the urethral meatus. A mean of 2.5 cm. was reported [56]. Marked variation has been recorded in the length of the clitoral body (glans and shaft). Frequently overlooked has been the possibility of an endocrine source for instances of hypertrophy of the organ observed clinically. Exact descriptions of points of origin of the clitoral crura on the anterior border of the symphysis or of any constant relation of crural origin to urethral meatus are an anatomic impossibility.

Aside from academic interest, Dickinson's expressed purpose in accumulating these data was to encourage clinical attempts to establish the physiology of clitoral function in female sexual response. In order to amplify his pioneer efforts, certain fundamental questions of clitoral reaction to sexual stimuli must be answered: (1) What anatomic changes occur in the clitoris during periods of sexual stimulation? (2) Are there consistent physiologic patterns of clitoral response that can be related to the descriptive framework of the four phases of the cycle of sexual response? (3) Does the clitoral body develop different response patterns during coition as opposed to those resulting from manipulation of the mons or other erogenous areas or to pure psychogenic stimulation? (4) What clinical application can be developed from the basic material accumulated to answer the first three questions? (5) Are clitoral and vaginal orgasms truly separate anatomic and physiologic entities? The questions relating to anatomy and physiology (Nos. 1–3) are approached immediately following, and the clinical questions (Nos. 4 and 5) are approached in Part 2 of this chapter.

The first two questions will be explored in sequence in order to define clitoral anatomic reaction and physiologic response to sexual stimuli within the descriptive framework of the four phases of the female cycle of sexual response [212, 214].

It should be reemphasized that there normally is marked variation in the anatomic structuring of the clitoris. Clitoral glandes frequently have been measured at 2 to 3 mm. in transverse diameter, yet a glans measuring 1 cm. in transverse diameter is still within normal anatomic limits. There also is marked variation in points of origin of the crural and suspensory ligaments. These ligaments originate on the anterior surface of the symphysis but vary from the lower to the upper border. The clitoral shaft (crura and corpora) may be quite long and thin and surmounted by a relatively small-sized glans, or short and thick with an enlarged glans. Frequently the reverse of these shaft-and-glans combinations has been observed. Clinical mensuration of clitoral shaft length has been so unreliable that results will not be reported.

The first pelvic response to sexual stimulation is the production of vaginal lubrication (see Part 1 of Chapter 6). This material appears on the walls of the vaginal barrel within 10 to 30 seconds from the onset of any form of sexual stimulation. Clitoral reaction does not develop as rapidly as the production of vaginal lubrication. Consequently the widespread belief that the clitoris responds to sexual stimulation with a rapidity equal to that of penile erection is fallacious. This physiologic misconception may have developed from the realization that anatomically the clitoris is a true homologue of the penis. It was a natural error to assume that similar anatomic structures would demonstrate parallel response patterns in a relatively equal time sequence.

The rapidity of clitoral response depends upon whether the stimulative approach is direct or indirect. The only direct approach is manipulation of the clitoral body or the mons area. There are numerous indirect stimulative techniques: manipulation of other erotic areas, coition, fantasy. If, for example, only breast or vaginal stimulation is employed (without direct clitoral contact), clitoral response will follow established patterns. However, there is a distinct delay in the onset of these patterns as opposed to the rapid reaction developed from direct stimulation of the clitoral body or the mons area.

EXCITEMENT PHASE

There is a clitoral response to sexual stimulation which occurs in every responding female during the excitement phase (Fig. 5-2) regardless of whether there is clinically obvious tumescence of the glans. The superficial integument of the unstimulated clitoral glans is wrinkled and moves without restriction over the underlying glans tissue in manner similar to the integument of the unstimulated glans penis, but with less freedom than the integument of the scrotum (see Chapter 13). When any form of sexual tension develops, the clitoral glans always increases in size to a

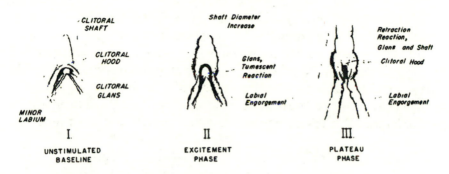

FIGURE 5-2

The clitoris in the female sexual response cycle. The orgasmic phase is omitted because of lack of information.

degree sufficient to develop close apposition between the subjacent tissues and the loosely applied, superficial integument. The vasocongestive reaction is of such finite nature that it usually cannot be noted by unsupported clinical observation. This anatomic response to increasing sexual tension has been established with aid of colposcopic magnification (6–40✕). Microscopic tumescence of the clitoral glans always develops with sexual tension, regardless of whether this vasocongestive process continues into a clinically observable (macroscopic) tumescent reaction.

There is no way of anticipating from observation in an unstimulated state whether or not a clitoral glans will develop a clinically obvious tumescence under sexual influence. When increase in size of the glans does occur, this reaction pattern develops with total consistency. Variations in tumescent reactions relate only to the rapidity and extent of increase in size of the glans in response to direct manipulation, as opposed to slower and less extensive glans tumescence in response to breast manipulation, active coition, or fantasy.

More than half of the study subjects did not develop clinically obvious tumescence of the clitoral glans. When macroscopic tumescence does occur, the degree of vasocongestion ranges from a barely discernible increase in diameter to a twofold expansion of the glans. This tumescent reaction of the glans has been confused with the penile erective process and has been mistermed "erection of the clitoris" [58, 97]. Total clitoral-body erection has not been observed unless there has been an obvious pathologic hypertrophy of the organ in its unstimulated state. Generally, the smaller the clitoral glans, the less frequently there is a clinically demonstrable tumescent reaction. However, some of the smaller organs have demonstrated the greatest relative size increases, while many of the larger clitorides have provided no gross evidence of a tumescent reaction.

When observable tumescence of the glans occurs, it does not develop until sexual tensions have progressed well into the excitement phase of the sexual response cycle. The clitoris engorges in a time sequence that parallels that of vasocongestion of the minor labia. It may be recalled that the minor labia of the sexually responding human female increase in size to a minimum of twice their unstimulated diameter and provide external extension for the expanding vaginal barrel (see Chapter 4). A similarly responding male has long since achieved full penile erection and, quite possibly, a moderate degree of elevation of at least one testicle (see Chapter 13).

Once observable tumescence of the clitoral glans develops, the engorgement persists throughout the remainder of the sexual cycle, or for as long as any significant degree of sexual stimulation is maintained.

As the anatomic structuring of the corpora cavernosa would suggest, the shaft of the clitoris also undergoes an excitement-phase vasocongestive reaction. There is definitive increase in diameter of the shaft which is a constant development regardless of shaft size. The vasocongestive increase in shaft diameter occurs simultaneously with the development of any macroscopic tumescent reaction of the glans. However, the clitoral shaft increases in diameter whether or not the glans reacts with clinically obvious tumescence.

In addition to a constant diameter increase, shaft elongation can occur. However, most clitorides go through vasocongestive glans reactions without developing clinically observable shaft elongation. Although objective observation admittedly is very difficult, an elongation reaction of the clitoral shaft has been firmly established in less than 10 percent of the observed orgasmic cycles. Shaft elongation develops only after the normal vasocongestive increase in shaft diameter has been stabilized. Elongation of the shaft has been observed only during direct manipulation of the mons area and not in response to the stimulation of other erotic areas of the body, fantasy, or active coition. It also should be emphasized that shaft elongation is confined to excitement-phase levels of sexual response, as discussion of plateau-phase clitoral response patterns will make evident.

PLATEAU PHASE

The most significant physiologic reaction of the clitoris to effective sexual stimulation occurs in the plateau phase of the sexual cycle (see Figs. 5-1, 5-2) and develops with universal consistency. The entire clitoral body (shaft and glans) retracts from the normal pudendal overhang positioning. The crura and suspensory ligaments of the clitoris have major anatomic functions in this pattern of physiologic response. The ischiocavernosus muscles also contribute actively to retraction of the clitoral body, as opposed to their function during male ejaculation (see Chapter 14). It should be emphasized that the exact roles of the crura, suspensory ligaments, and various muscle bundles in clitoral retraction have not been determined with total conviction.

Clitoral reaction to plateau-phase levels of sexual tension occurs in a constant pattern. The shaft and glans of the clitoris withdraw from normal pudendal-overhang positioning and retract against the anterior border of the symphysis. Any portion of the clitoral glans that normally projects from the clitoral hood in a sexually unstimulated state is withdrawn deeply beneath the protective foreskin as the retraction reaction progresses. In the immediate preorgasmic period the clitoral body (shaft and glans) is extremely difficult to observe clinically. At this time the retraction of the clitoral shaft normally is so advanced that there is at least a 50 percent overall reduction in the length of the total clitoral body. The degree of individual clitoral-body retraction has been estimated with the aid of direct colposcopic observation.

During the plateau phase, clitoral-body retraction develops in relation to mode and effectiveness of sexual stimulation. During coition or breast manipulation, clitoral retraction develops late in the plateau phase as an indication of preorgasmic levels of sexual tension. With manipulation of the mons area, retraction of the clitoral shaft develops more rapidly, frequently early in the plateau phase, and may indicate sensate response to tactile stimuli rather than imminence of orgasmic experience.

Obviously, psychic components of sexual response patterns cannot be equated objectively for each orgasmic cycle under investigation. However, it would be a major mistake to presume that psychogenic influences do not contribute to either degree or rapidity of clitoral response to effective sexual stimulation. No woman who can fantasy to advanced plateau stages of sexual tension has been available to the investigation, so there is no information as to the degree of clitoral retraction possible in response to purely psychosexual stimulation.

Retraction of the clitoral body during the plateau phase is a reversible reaction. If high sexual tension levels are allowed to fall by deliberate reduction or withdrawal of stimulative techniques, the retracted shaft and glans will return to the normal pudendal-overhang position. With return to effective sexual stimulation, clitoral-body retraction will recur. This clitoral reaction sequence may develop repetitively during long-maintained plateau phases. Such a situation would exist when a woman who cannot quite achieve orgasmic expression insists on long-continued or repetitive return to stimulative activity in attempts to obtain release from her demanding sexual tensions.

ORGASMIC PHASE

No specific orgasmic-phase reaction of the clitoris has been established. In fact, due to the severity of the normal clitoral retraction beneath the minor labial hood, the clitoral glans has never been available to direct observation during an orgasmic experience.

RESOLUTION PHASE

After an orgasmic episode the return of the clitoris to normal pudendal-overhang positioning occurs within 5 to 10 seconds after cessation of orgasmic platform contractions (see Part 1 of Chapter

6). Retraction of the clitoral body is reversed even more rapidly than detumescence of the orgasmic platform and as swiftly as the sex-skin discoloration disappears from the minor labia (see Chapter 4). To provide further concept of this rapid clitoral-body "release," a parallel might be drawn to the male reaction pattern. The relaxation of the retracted clitoral shaft and the return of the glans to the normal pudendal-overhang positioning occurs in a parallel time sequence with the primary-stage involution of male penile erection after ejaculation (see Part 1 of Chapter 12).

When an observable tumescence of the clitoral glans has developed during the excitement phase, subsequent resolution-phase detumescence of the glans is a relatively slow process. This is particularly true for the individuals who demonstrate as much as a twofold vasocongestive increase in glans size. Although termination of the clitoral retraction reaction occurs very rapidly, continued tumescence of the glans and vasocongestion of the shaft frequently have been observed to last 5 to 10 minutes after orgasmic expression. Occasionally some women have demonstrated continued venous engorgement of the clitoral shaft or glans for 15 to 30 minutes after an orgasmic experience.

Those individuals who achieve plateau-phase levels of sexual response but do not obtain orgasmic-phase release of the accumulated sexual tensions occasionally maintain venous engorgement of both clitoral shaft and glans for a matter of hours after termination of all sexually stimulative activity.

PATTERNS OF CLITORAL RESPONSE

With the answers to the first two questions established and available as anatomic and physiologic baselines, the third question may be approached with more security. Does the clitoris develop different response patterns during coition as opposed to manipulation of the mons or other erogenous areas or to pure psychogenic stimulation?

Clitoral response was observed during natural coital activity in three positions: female supine, superior, and knee-chest; during artificial coition in female supine and knee-chest positions; and during both manual and mechanical manipulation of the mons and other erogenous areas. In female supine position, during natural coition, it was impossible to establish accurate observations of clitoral reaction patterns. The information returned from female supine positioning was developed by artificial coital techniques. Conversely, artificial coition was a technical impossibility in female superior coital positioning, so information of clitoral-body reaction patterns in this position was returned only from natural coital activity. Regardless of positioning and type of coition, or erogenous areas manipulated, the reactions of the clitoris to successful sexual stimulation followed the physiologic response patterns detailed in preceding portions of this chapter.

When women developed clinically obvious tumescence of the clitoral glans subsequent to mons manipulation, they achieved similar degrees of vasocongestion during coition in the three described positions. Conversely, none of the female study subjects developed glans tumescence during coition without demonstrating similar or more severe degrees of glans vasocongestion during mons area stimulation. When the mons was manipulated directly, the observable tumescent reaction of the glans occurred earlier in the excitement phase than when this vasocongestive reaction developed during coital activity.

Vasocongestive increase in diameter of the clitoral shaft occurs in all women regardless of the presence or absence of an observable glans tumescence. This reaction developed earlier in the excitement phase when direct mons area manipulation was employed than when the study subjects were responding to coital stimulation. The presence or absence of a clitoral shaft-elongation reaction could not be determined during active coition.

Three women were able to achieve orgasmic response by breast manipulation alone, in addition to their ability to react with orgasmic success to mons manipulation and to coition. Only one of the three women demonstrated an observable excitement-phase tumescent reaction of the clitoral glans during mons manipulation, coition, or breast stimulation. There was obvious delay in this secondary vasocongestive reaction during breast automanipulation when compared with the rapid development of glans tumescence during direct manipulation of the mons area.

As anticipated, the retraction reaction of the clitoral body (glans and shaft) developed during the plateau phase for all three breast manipulators. The reaction paralleled in time sequence that achieved during intercourse (late plateau, preorgasmic phase) and obviously was delayed as compared to the more rapid response patterns (early plateau phase) elicited by direct mons manipulation.

Unfortunately, as mentioned earlier, study subjects available to the investigation did not include individuals who could fantasy to orgasm. Therefore, observations of clitoral-body reaction patterns subsequent to psychogenic sexual stimulation have been limited to excitement-phase levels of sexual response. This level of sexual tension has been created frequently by providing suggestive literature for the study subjects. A clinically obvious tumescent reaction of the clitoral glans could be demonstrated in only a few of the women who normally developed this reaction during somatogenic stimulation. When clinical tumescence of the glans did occur, it developed long after there was obvious production of vaginal lubrication and paralleled a vasocongestive increase in the size of the minor labia. A minimum of a half-hour of exposure to stimulative literature was necessary to produce an observable glans tumescence in any woman.

The microscopic vasocongestive reaction of the clitoral glans which provides close apposition between integument and underlying glans tissues occurred in approximately 75 percent of the women who were exposed to suggestive literature. Fewer than one-third of the responding women produced a demonstrable increase in clitoral shaft diameter and no shaft-elongation reaction was observed.

The only variations in clitoral response developed from psychogenic forms of sexual stimulation, natural or artificial coition, and manual or mechanical manipulation of the mons or other erogenous areas have been in the rapidity and intensity of physiologic reaction.

2 . CLINICAL CONSIDERATIONS

Clinical error has dominated the assignment of clitoral function in sex tension increment for the human female. Therefore, a detailed consideration of the dual capacity of the clitoris, as both a receptor and transformer of sexual stimulation, is in order. The definitive role of the clitoris in sexual response must be appreciated if female sexual inadequacy ever is to be treated effectively.

Five questions have been raised and must be answered if the role of the clitoris in human female sexuality is to be established. Three of these five questions have been discussed in Part 1 of this chapter. Two questions remain: (4) What clinical application can be developed from the basic material accumulated to answer the first three questions? (5) Are clitoral and vaginal orgasms truly separate anatomic and physiologic entities? Attempts to answer these two questions have directed investigative interest toward a clinical evaluation of clitoral influence upon female sexual response.

While the literature contains innumerable discussions of the role of the clitoris in female sexuality, authoritative opinion has reached essential accord only in the view that the primary function of the organ is to stimulate female sexual tensions. In order to accomplish its clinical purpose, the clitoris functions in the dual capacity of both receptor and transformer of sexual stimuli regardless of whether these stimuli originally have been somatogenically or psychogenically oriented. This concept will be discussed later.

In the past, attempts have been made to assign to clinical variants in clitoral anatomy and physiology specific influence on the total of female sexual response. Eleven years of investigation have failed to support these concepts. Both the size of the clitoral glans and the total clitoral body's positioning on the anterior border of the symphysis have been assigned roles of major influence in female sexual response [8, 144, 268, 319, 330]. However, Dickinson and Pierson [54, 59] originally expressed the conviction that there is no relation between the size of the clitoris and the effectiveness of its role in female sexual stimulation. Direct observation of thousands of sexual response cycles has confirmed their opinion.

Historically, the anatomically oriented concept that clitoral size has a direct relation to the effectiveness of the individual female's sexual performance has been fostered by our "phallic fallacy" literature and has no foundation in fact. The diameter of the unstimulated clitoral glans measured at the juncture of the glans and shaft has varied in the study-subject population from 3 mm. to 1 cm. In this group there has been absolutely no relationship established between the size of the glans and the rapidity and intensity of the individual's ability to respond to effective sexual stimulation. Conversely, there also has been similar variation in clitoral glans size among women treated for inadequacy of sexual response during the past seven years. Regardless of the type of somatogenic or psychogenic approach to sexual stimulation, clitoral glans size has played no definitive part in the effectiveness of the individual's sexuality.

Consideration of the anatomic positioning of the clitoris has paralleled any discussion of clitoral size in relation to degree of sexual response. Clitoral placement on the anterior border of the symphysis has been assigned a role of major influence on female response during coition [57, 319, 330]. A low implantation has been presumed to improve the sexuality of the individual female due to the possibility of increased direct contact between the penis and clitoral glans. Regardless of clitoral-body positioning the penis rarely comes in direct contact with the clitoral glans during active coition. In fact, clitoral retraction, which always develops during the plateau phase and elevates the clitoral body from its normal pudendal-overhang positioning, further removes the glans from even the theoretical possibility of direct penile contact.

Specific physiologic reactions, like anatomic variants of the clitoris, also have been assigned major roles in elevating female sexual tensions [8, 56, 64, 65, 140, 144, 305, 306, 319, 330]. Studies in depth of both the study-subject population and women undergoing treatment for sexual inadequacy have failed to support these contentions. Whether the clitoris develops an obvious tumescence of the glans or elongation of the shaft has little to do with the degree of the individual woman's response to effective sexual stimulation. Tumescence of the glans and shaft elongation have been observed in women during multiple orgasmic sessions as well as in women who have not been able to achieve orgasmic levels of sexual tension. Clitoral-body retraction occurs during the plateau phase whether or not an orgasmic experience is to follow. In brief, sexually responding women achieve orgasmic levels of sexual tension without regard to variables in the basic anatomy and physiology of the clitoris.

Dickinson [56] insisted that women with histories of decades of masturbatory activity did not develop a consistent hypertrophy of the clitoris. While this general concept certainly is acceptable, there are minor exceptions that should be noted. Observations of individuals over the past decade have removed any doubt that frequent, severe masturbatory activity occasionally may produce measurable increases in the diameter of the clitoral glans and questionable increases in the length of the clitoral shaft. When recordable clitoral glans hypertrophy develops over a period of years, the women usually are found to employ extensively one or more of the mechanical methods for clitoral stimulation. Obviously, long-continued androgenic influence (adrenal hyperplasia, testosterone ingestion, etc.) must be ruled out first in these cases.

It may be recalled that there are reports of African tribes that measure female sexuality in terms of clitoral length and labial hypertrophy [249]. From infancy, female members of such tribes deliberately are manipulated for countless hours to stimulate the development of these artifacts. These girls have been reported to obtain an obvious hypertrophy of the clitoris and the labia, if not by puberty, at least during their early teens. Although the fact of manipulative hypertrophy is established, there is no reliable information relating the hypertrophy directly to excessive levels of female sexuality. It is possible that methods used to attain a culturally desirable condition of adornment can simultaneously increase individual sexual responsiveness.

Although anatomic placement and physiologic reaction preclude any consistency of direct clitoral glans stimulation during coition, the significant influence of secondary stimulation should not be overlooked. The fact that the clitoral glans rarely is contacted directly by the penis in intravaginal thrusting does not preclude the coital development of indirect clitoral involvement. Clitoral stimulation during coitus in the female supine position develops indirectly from penile-shaft distention of the minor labia at the vaginal vestibule. A mechanical traction develops on both sides of the clitoral hood of the minor labia subsequent to penile distention of the vaginal outlet. With active penile thrusting, the clitoral body is pulled downward toward the pudendum by traction exerted on the wings of the clitoral hood. However, there is not sufficient excursion developed by coital traction on the clitoral body to allow direct penis-to-clitoris contact.

When the penile shaft is in the withdrawal phase of active coital stroking, traction on the clitoral hood is somewhat relieved and the body and glans return to normal pudendal-overhang positioning. However, the rhythmic movement of the clitoral body in conjunction with active penile stroking produces significant indirect or secondary clitoral stimulation.

It should be emphasized that this same type of secondary clitoral stimulation occurs in every coital position when there is a full penetration of the vaginal barrel by the erect penis. Anatomic exceptions to this statement are created by any significant pathologic gaping of the vaginal outlet, such as might be occasioned by childbirth injury. If the vaginal outlet is too expanded to allow strong traction on the minor-labial hood by the thrusting penis, minimal clitoral excursion will occur and little if any secondary stimulation will develop.

Only the female superior and lateral coital positions allow direct or primary stimulation of the clitoris to be achieved with ease. In these positions the clitoris can be stimulated directly if apposition between male and female symphyses is maintained. There also remains the constant factor of secondary clitoral stimulation provided by traction on the minor-labial hood during active coition in these positions. The influences of both direct and indirect stimulation are essentially inseparable in these coital positions. Clitoral response may develop more rapidly and with greater intensity in female superior coition than in any other female coital position.

In the knee-chest coital position no direct stimulation of the clitoris is possible. Yet glans tumescence, when it occurs, and clitoral-body retraction, which is a constant factor, occur in the response patterns established for the supine or superior coital positions. The intensity of physiologic reaction usually is less pronounced than in either supine or superior coital positioning.

Obviously, active coition develops psychogenic as well as physiologic response patterns, both of which contribute to indirect or secondary clitoral stimulation. It will remain for more sophisticated methods of neurophysiologic and psychologic investigation to assign individual spheres of influence to these multiple influences which create the total picture of indirect stimulation of the receptor organ developed by active coition.

In essence, stimulation of the clitoris (receptor organ) developing during active coition is the secondary or indirect result of penile traction on the minor labial hood. This traction occurs regardless of female coital positioning, anatomic variations in clitoral size, or crural origin on the pubic rami.

The importance of development by marital units of specific coital techniques to facilitate clitoral stimulation has been emphasized repeatedly in the literature [15, 68, 144, 150, 163, 193, 278, 305, 319, 330]. The clinical fallibility of these suggestions now is obvious. Unless the male partner makes a specific effort to bring the shaft of the penis in direct apposition to the total mons area, the clitoris is not stimulated directly by penile thrust with the female in the usual supine position. An overriding coital position is difficult for the male partner to maintain as sexual tensions increase, particularly if the female does not have parous relaxation of the vaginal outlet. The nulliparous woman may not be able to retain the penis in an awkward pelvic override position without complaining of vaginal outlet or rectal discomfort.

An additional objection to the male-override position is that it precludes full vaginal penetration at the apex of the penile thrust. Thus the mutual coital stimulation of vaginal engulfment for the male and cul-de-sac distention for the female are lost to the sexual partners. Intensity of vaginal exteroceptive and proprioceptive response can be dulled for the female partner by any awkward attempt to provide direct clitoral glans contact.

The primary focus for sensual response in the human female's pelvis is the clitoral body. The clitoris responds with equal facility to both somatogenic and psychogenic forms of stimulation, and is truly unique in the human organ system in that its only known function is that of serving as an erotic focus for both afferent and efferent forms of sexual stimulation. How, then, does the clitoral body function in its role as receptor and transformer to sexually invested stimuli?

At the outset it should be made perfectly clear that although stimuli are characterized as somatogenic or psychogenic in origin and the roles of the clitoris as receptor and transformer, this does not imply that any form of stimulation is or can be purely somatogenic in character. All stimuli are appreciated, delineated, and referred by higher cortical centers. The term *somatogenic* relates only to physical activity. This form of clitoral stimulation can vary from heterosexual manual manipulation to automanipulative use of bedding material or thigh pressure. Thus the use of the terms *somatogenic stimuli* or *transformer role* connotes initiation or approach rather than any concept of discriminatory ability.

Sexual stimuli may be derived from either somatogenic or psychogenic origins. The clitoral response patterns will vary depending upon the initial involvement of either afferent or efferent pathways. When the clitoral body reacts directly to automanipulative techniques or secondarily to coital activity, these stimuli (initially somatogenic but with an obvious psychogenic overlay) are received through the afferent nerve endings in the clitoral glans and shaft. Clitoral-body response to this type of stimulation could, from a clinical point of view, be termed *receptor* in character.

The pacinian corpuscles within the large nerve bundles conceivably play an important role in relaying afferent impulses created by somatogenic forms of stimulation. As Krantz [156] so ably has shown, there is marked variation in quantity and quality of nerve endings and in the number of pacinian corpuscles located within the individual clitoral glans and shaft. Since the assigned role of the pacinian corpuscles is that of proprioceptive response to deep pressure (receptor role), the great variety in female automanipulative techniques ranging from demand for severe pressure to insistence upon the lightest touch may be explained.

Little is known of the neurologic pathways that lead from

stimulated afferent nerve endings in the clitoral body. Although a reflex center in the sacral portion of the spinal cord has been identified in the male animal by Semans and Langworthy [285], no similar response center has been described for the human female. It may be that the entire reflex arc involving the spinal cord and the higher cortical centers constantly is caught up in the continuum of response to dominantly somatogenic forms of sexual stimulation. Particularly is this concept plausible when it is realized that regardless of the effectiveness of the somatogenically oriented stimuli, the psychogenic overlay inherent in any approach to female sexual stimulation is of constant import. Therefore, the possibility of a pure reflex-arc response to afferent stimulation is reduced with the realization that psychogenic stimulation of the higher cortical centers and the resultant direct, efferent, transformer response in the clitoris is an undeniable factor in the sexual response of the human female.

The clitoral body functions as a receptor organ in an objective expression of sensual focus, as well as the subjective end-point (transformer) of neurogenic pathways. The result of efferent stimulation of the clitoris, be it psychogenically or somatogenically initiated, has been recorded in the detailed consideration of the anatomy and physiology of the clitoral body's response to varying intensity of sexual stimulation (see Part 1 of this chapter). However, the functional role (that of serving clinically as a transformer or subjective organ of sensual focus) has not been considered previously.

The subjective, or transformer, response of the clitoris to any form of effective sexual stimulation, such as reading of pornography, direct manipulation, coital connection, etc., has been vocalized by women in many ways. Some vocally identify a subjective sensation of deep pelvic fullness and warmth (possibly vasoconcentration), others a feeling of local irritation, expansive urge, need for release, etc. (possibly glans enlargement). The clinical or functional response of the clitoris as a transformer of efferent forms of stimulation is to create in turn a subjective urge or tension increment and, ultimately, a higher cortical need for release. It is impossible to delimit this functional clitoral role of sensual focus because vocalization of the sensual response patterns varies from woman to woman. The transformer role also differs between clitoris and penis (see Part 2 of Chapter 12). Suffice it to say that the clitoris, serving as a receptor and transformer organ, has a role as the center of female sensual focus, and the functional response it creates easily is identifiable by any sexually oriented woman.

Any clinical consideration of clitoral response to effective sexual stimulation must include a discussion of masturbation. The techniques of and reactions to direct manipulation of the clitoral body (glans and shaft) or the mons area vary in each woman. Observations of higher animal patterns of foreplay first sensitized investigators to the clinical importance of effective autostimulative techniques by emphasizing the obvious response that such effective foreplay can develop in the female of the species [10, 76].

Marriage manuals discuss at length the importance of clitoral manipulation as the basis of adequate coital foreplay. Most discussions of initiation and elevation of female sexual tensions have included the questions of why and when to stimulate the clitoris. To date there has been little consideration of the infinitely more important questions of how to manipulate the clitoris and how much stimulation usually is required. Direct observation of hundreds of women using mechanical and manual masturbatory techniques through repetitive orgasmic experiences has emphasized the fundamental importance of the questions, "How?" and "How much?"

No two women have been observed to masturbate in identical fashion. However, there is one facet of general agreement. Women rarely report or have been noted to employ direct manipulation of the clitoral glans. In those isolated instances when the technique is used it is limited to the excitement phase only and frequently a lubricant is applied to this normally quite sensitive tissue. Additionally, the clitoral glans often becomes extremely sensitive to touch or pressure immediately after an orgasmic experience, and particular care is taken to avoid direct glans contact when restimulation is desired.

Those women who manipulate the clitoris directly concentrate on the clitoral shaft. Usually they manipulate the right side of the shaft if right handed, and the left side if left handed. Occasionally, women have been observed to switch sides of the shaft during stimulative episodes. A relative degree of local anesthesia may develop if manipulation is concentrated in just one area for extended periods of time or if too much manipulative pressure is applied to any one area.

Women usually stimulate the entire mons area rather than concentrating on the clitoral body. Regardless of whether the clitoris is stimulated by direct means or indirectly through mons area manipulation, the physiologic responses of the clitoris to elevated sexual tensions are identical. Most women prefer to avoid the overwhelming intensity of sensual focus that may develop from direct clitoral contact. Instead, mons area manipulation produces a sensual experience that although somewhat slower to develop is, at orgasmic maturity, fully as satiating an experience as that resulting from direct clitoral shaft massage. Mons area manipulation also avoids the painful stimuli returned to many women when the clitoris is manipulated directly either with too much pressure or for too lengthy periods of time.

The concept of the mons as an area of severe sensual focus is supported by the clinical observation that after clitoridectomy, masturbation has been reported to be as effective a means of sexual stimulation as before surgery [23]. Manipulation usually has been confined to the mons area, although sometimes concentrated on the scarred postsurgical site.

Evidence of the extreme tactile sensitivity of the entire perineum in addition to the clitoral body and the mons area has been presented by the Institute for Sex Research [144]. During the Institute's gynecologic observation, the minor labia were determined to be almost as perceptive to superficial tactile sensation as the clitoral glans. The Institute also considers the minor labia to be fully as important as the clitoris or mons as a source of erotic arousal. While the tactile sensitivity of the minor labia is without question, stimulation of the labia does not provide the human female with the extremes of sensual stimuli that massage of the clitoral shaft or mons area produces.

Another observation of female automanipulative technique should be considered for its clinical import. Most women continue active manipulation of the clitoral shaft or mons area during their.

entire orgasmic experience. This female reaction pattern parallels their coital pattern of demand for continued active male pelvic thrusting during the woman's orgasmic experience. This female demand for continued stimulation during the actual orgasmic expression is in opposition to the average male's reaction to his ejaculatory experience. Most males attempt the deepest possible vaginal penetration as the first stage of the ejaculatory response develops. They maintain this spastic, deep vaginal entrenchment during the second phase of the ejaculatory experience rather than continuing the rapid pelvic thrusting characteristic of preorgasmic levels of sexual tension (see Part 2 of Chapter 12).

The human female frequently is not content with one orgasmic experience during episodes of automanipulation involving the clitoral body. If there is no psychosocial distraction to repress sexual tensions, many well-adjusted women enjoy a minimum of three or four orgasmic experiences before they reach apparent satiation. Masturbating women concentrating only on their own sexual demands, without the psychic distractions of a coital partner, may enjoy many sequential orgasmic experiences without allowing their sexual tensions to resolve below plateau-phase levels. Usually physical exhaustion alone terminates such an active masturbatory session.

There is a specific clitoral-body reaction to effective sexual stimulation that has created a state of confusion for the average male sexual partner. This physiologic response to sexual tension has been termed the *retraction reaction*. The entire clitoral body is elevated high on the anterior border of the symphysis (away from its normal pudendal-overhang positioning) during both the plateau and orgasmic phases of the female sexual response cycle (see Part 1 of this chapter).

This physiologic reaction to high levels of female sexual tension creates a problem for the sexually inexperienced male. The clitoral-body retraction reaction frequently causes even an experienced male to lose manual contact with the organ. Having lost contact, the male partner usually ceases active stimulation of the general mons area and attempts manually to relocate the clitoral body. During this "textbook" approach, marked sexual frustration may develop in a highly excited female partner. By the time the clitoral shaft has been relocated, plateau-phase tension levels may have been lost. Not infrequently the female partner, frustrated by male ineptitude, may not recover from her psychophysiologic distraction sufficiently to avoid the frustrating, vasocongestive pelvic distress occasioned by orgasmic inadequacy.

It is important to reemphasize the fact that the retracted clitoral body continues to be stimulated by traction or pressure on the protective clitoral hood. Once plateau-phase clitoral retraction has been established, manipulation of the general mons area is all that is necessary for effective clitoral-body stimulation.

Most marriage manuals advocate the technique of finding the clitoris and remaining in direct manual contact with it during attempts to stimulate female sexual tensions. In direct manipulation of the clitoris there is a narrow margin between stimulation and irritation. If the unsuspecting male partner adheres strictly to marriage manual dictum, he is placed in a most disadvantageous position. He is attempting proficiency with a technique that most women reject during their own automanipulative experiences.

As stated previously, no two women practice automanipulation in similar fashion. Rather than following any preconceived plan for stimulating his sexual partner, the male will be infinitely more effective if he encourages vocalization on her part. The individual woman knows best the areas of her strongest sensual focus and the rapidity and intensity of manipulative technique that provides her with the greatest degree of sexual stimulation.

Finally, a brief consideration of the fifth and last of the questions raised about the role of the clitoris in female sexuality: Are clitoral and vaginal orgasms truly separate anatomic entities? From a biologic point of view, the answer to this question is an unequivocal No. The literature abounds with descriptions and discussions of vaginal as opposed to clitoral orgasms [14, 24, 42, 53, 63, 83, 84, 115, 116, 153, 159, 177, 277, 324, 331]. From an anatomic point of view, there is absolutely no difference in the responses of the pelvic viscera to effective sexual stimulation, regardless of whether the stimulation occurs as a result of clitoral-body or mons area manipulation, natural or artificial coition, or, for that matter, specific stimulation of any other erogenous area of the female body.

With the introduction of artificial coital techniques, the reactions of the vagina during coition became available to direct observation and repeatedly have been recorded through the medium of cinematography. These vaginal reactions first had been observed during sexual response cycles stimulated by manipulation of the mons area and clitoral body [203, 205, 207]. During artificial coition the reactions of the vaginal barrel initiated under direct stimulation conformed in exact detail to the vaginal response patterns which developed subsequent to the indirect stimulation of mons area or clitoral-body manipulation.

Three study subjects available to the investigative program have demonstrated the facility of orgasmic response to breast stimulation alone, as well as to coital, clitoral-body, or mons area manipulation. Identical vaginal response patterns were observed for these three study subjects from all the above-described modes of stimulation.

Conversely, what of clitoral-body reaction to direct or indirect stimulation? The physiologic responses that develop in the clitoral glans and shaft during the four phases of the sexual cycle are the same regardless of whether the clitoral body is responding to direct or indirect stimulation. For research purposes the definition of indirect or direct clitoral-body stimulation has been oriented to clinical considerations alone.

Direct stimulation results from manual or mechanical manipulation of the clitoral shaft or glans. Indirect stimulation develops from mons area manipulation or the stimulation of any other erogenous area of the female body, such as the breasts. In addition, the clitoral body may be stimulated indirectly by natural or artificial coition with the female partner in the supine, superior, or knee-chest position. All of these techniques have been used in order to record clitoral-body response patterns. These patterns are identical and vary only in intensity of reaction to the effectiveness of the stimulative technique, regardless of whether this technique is described clinically as direct or indirect.

There may be great variation in duration and intensity of orgasmic experience, varying from individual to individual and within the same woman from time to time. However, when any woman experiences orgasmic response to effective sexual stimulation, the vagina and clitoris react in consistent physiologic patterns. Thus, clitoral and vaginal orgasms are not separate biologic entities.

6

THE VAGINA

1. ANATOMY AND PHYSIOLOGY

The vaginal barrel performs a dual role, providing the primary physical means of heterosexual expression for the human female and serving simultaneously as an integral part of her conceptive mechanism. Discussion in this part of the chapter will reflect the anatomic and physiologic reactions of the vagina in its role as the physical means by which the woman expresses sexual capacity. A consideration of the efficiency of the vagina in conceptive physiology will be presented in Parts 2 and 3 of this chapter. The duality of the vaginal role should be emphasized. It is possible that the efficiency of vaginal conceptive function is a physiologic measure of the psychosomatic effectiveness of the vagina's role as the primary means of the female's sexual expression.

To appreciate vaginal anatomy and physiology is to comprehend the fundamentals of the human female's primary means of sexual expression. In essence, the vaginal barrel responds to effective sexual stimulation by involuntary preparation for penile penetration. Just as penile erection is a direct physiologic expression of a psychologic demand to mount, so expansion and lubrication of the vaginal barrel provides direct physiologic indication of an obvious psychologic mounting invitation.

When the role of the vagina as a sexual organ is evaluated, the physiologic responses of the vaginal barrel to sex tension increment come into primary focus. Indeed, the reactions of the artificial as well as the natural vaginal barrel must be considered (see Chapter 7). The anatomic alterations that develop within the natural or artificial vaginal barrel, when it is functioning as the primary means of female sexual expression, frequently are oriented to specific levels of sexual tension. Therefore, these tension-induced anatomic variations will be described with relation to the four phases of the cycle of sexual response.

It should be stated parenthetically that vaginal (natural or artificial) response to sexual stimulation develops in a basic pattern regardless of whether the stimuli originally are primarily somatogenic or psychogenic in origin.

EXCITEMENT PHASE

The first physiologic evidence of the human female's response to any form of sexual stimulation is the production of vaginal lubrication. Lubricating material appears on the walls of the vagina within 10 to 30 seconds after the initiation of any form of effective sexual stimulation. There have been many efforts over the years to discover the source of vaginal lubrication. The cervix has been considered by many authors as the primary site of the lubricating material [8, 21, 30, 54, 67, 144, 191, 239, 295, 308, 318, 319]. Bartholin's glands (see Chapter 4) also have been allotted a major role in the production of vaginal lubrication. Neither the healthy cervix nor the Bartholin's glands make any essential contribution to the total of vaginal lubrication.

As sexual tensions rise, a "sweating" phenomenon may be observed developing on the walls of the vaginal barrel (Fig. 6-1). Individual droplets of transudation-like, mucoid material appear scattered throughout the rugal folds of the normal vaginal architecture. These individual droplets present a picture somewhat akin to that of the perspiration-beaded forehead. As tensions increase, the droplets coalesce to form a smooth, glistening coating for the entire vaginal barrel. This "sweating" phenomenon provides complete lubrication for the vaginal walls early in the excitement phase of the human female's sexual response cycle and certainly is the first evidence of the vaginal barrel's physiologic response to sexual stimulation. In a matter of seconds the sexually responding woman may develop sufficient lubrication for coital readiness.

Identification of the vagina's lubricating mechanism has been one of the most interesting aspects of the anatomic study of the

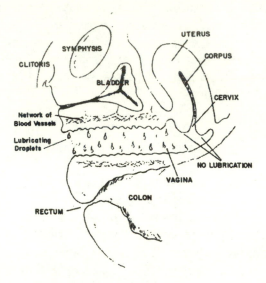

FIGURE 6-1

Schematic representation of vaginal lubrication.

human female's sexual response cycle. Present information suggests that this transudation-like material is the result of marked dilatation of the venous plexus which encircles the entire vaginal barrel. The bulbus vestibuli, plexus pudendalis, plexus uterovaginalis, and, questionably, the plexus vesicalis and the plexus rectalis externus are all involved in a fulminating vasocongestive reaction about the walls of the vagina. Apparently the transudation-like material which lubricates the vagina develops from the activation of a massive localized vasocongestive reaction. It long has been established that there are essentially no glandular elements in the walls of the vagina [219]. Yet this transudation-like material appears readily, usually in copious amounts, early in the excitement phase of the sexual response cycle. The biochemical constitution of the lubricating material and its effect upon normal vaginal chemistry are under present investigation.

In only one instance has any cervical activity of a secretory nature been noted during directly observed intravaginal cycles of sexual response. In this particular situation the study subject lost a mucus plug from the cervical outlet. She was in the thirteenth day of a regularly recurring 28-day menstrual cycle. It was interesting to note that although this thin, ovulatory-type mucus was extruded from the cervical canal, the reaction occurred late in the plateau phase of the sexual response cycle, long after vaginal lubrication had been established.

In a consideration of the problem of vaginal lubrication, there are two other items of information which should be presented. First, a pronounced sweating phenomenon has been repeatedly observed in women who have been subjected to complete hysterectomy and bilateral salpingo-oophorectomy. The ability to lubricate with reasonable effectiveness, even in a state of complete castration, mechanically eliminates the cervix from significant consideration as a primary source of lubrication. Second, still further verification that the cervix makes no contribution to vaginal lubrication is provided by those women for whom artificial vaginas have been created (see Chapter 7). They also produce an effective degree of vaginal lubrication by means of the transudation-like "sweating" mechanism.

In order to comprehend the distensive ability of the vaginal barrel in response to sexual stimuli, it must be recalled that anatomically the unstimulated vagina is a potential rather than an actual space. Unless the woman is menstruating or subjected to sexual stimulation, the anterior and posterior walls of the vagina are essentially contiguous. Obviously, the minimal spacing shown between the anterior and posterior vaginal walls in Figure 6-2, representing normal pelvic anatomy, is, in truth, inaccurate. Obstetricians are well aware that the normal vagina is infinitely distensible from a clinical point of view. The demands of childbirth clearly establish that the collapsed state of the unstimulated vaginal barrel is no measure of the vagina's expansive potential.

As the excitement phase of sexual tension continues, further anatomic evidence of psychologic mounting readiness develops within the vagina (Fig. 6-3). Initially, there is a lengthening and distention of the inner two-thirds of the vaginal barrel. Irregular expansive movements of the vaginal walls may be observed as sexual tensions mount toward the plateau phase. Before termination of the excitement phase, the vaginal barrel is markedly expanded.

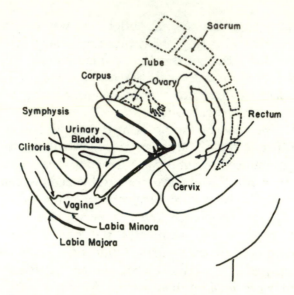

FIGURE 6-2
Female pelvis: normal anatomy (lateral view).

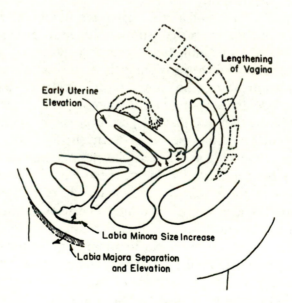

FIGURE 6-3
Female pelvis: excitement phase.

The walls of the inner two-thirds of the vaginal barrel expand involuntarily and then partially relax in an irregular, tensionless manner. Slowly the demand to expand overcomes the tendency to relax, and the clinically distended vaginal barrel of the sexually responding woman is established.

In addition to the expansive effect in the vaginal fornices, the cervix and corpus are pulled slowly back and up into the false pelvis as sexual tensions mount (see Chapter 8). This reaction, together with the expansion of the fornices, contributes greatly to the distention of the inner two-thirds of the vaginal barrel. Cervical elevation creates a tenting effect at the transcervical depth in the midvaginal plane. The slow, irregular elevation of the cervix from the vaginal axis toward the false pelvis is dependent upon normal anterior uterine positioning. If the uterus is in marked 3° retroversion, for example, not only is there no cervical elevation from the vaginal axis, but also the usual degree of mid-vaginal expansion and posterior vaginal extension is reduced.

The sexually unstimulated vaginal barrels (Fig. 6-4) of 100 nulliparous study subjects have been measured repeatedly at a 2 cm. diameter in the transcervical plane (just anterior to the resting cervix of an anteriorly placed uterus). As sexual tensions mount, transcervical vaginal-wall expansion ranges from 5.75 to

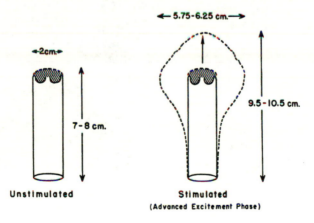

FIGURE 6-4

Nulliparous vaginal barrel: baseline measurements.

6.25 cm. Vaginal length (fourchette to posterior-fornix wall) of the unstimulated nulliparous vagina has been recorded as ranging from 7 to 8 cm. During excitement-phase response the vaginal length measurement increases to 9.5–10.5 cm.

In an attempt to demonstrate the vagina's essentially unlimited clinical distensibility, the same transcervical vaginal-wall and vaginal-length measurements were taken within the 100 nulliparous vaginas previously distended by an indwelling speculum (Fig. 6-5). The speculum was placed at a fixed dilation of 2 cm. between the anterior and posterior blades. The initial transcervical expansion readings were now an average of 3–4 cm., and vaginal length was recorded at an 8–9 cm. average. With excitement-phase response, the transcervical expansion of the vaginal walls reached averages of 6.75–7.25 cm., and vaginal-length extension was recorded at an 11–12 cm. average.

From these figures it is obvious that the more the vagina dilates, the more further dilatation becomes possible. When the two sets of figures are compared, a maximum increase of 3.75–4.25 cm. in transcervical width and 2.5–3.5 cm. in length have been established for the nulliparous vagina under the influence of excitement-phase

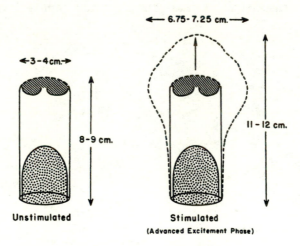

FIGURE 6-5

Nulliparous vaginal barrel: speculum-dilated.

levels of sexual tension. When the vagina is partially dilated by an indwelling speculum, the transcervical-diameter increase under the additional influence of effective sexual stimulation reaches a maximum of 3.75–4.25 cm., and the length increase averages 3–4 cm. From these essentially similar figures it is obvious that under sexual stimulation the inner two-thirds of the vagina expands and the vaginal barrel lengthens in complete disassociation from previously established states of vaginal distention.

This experiment was not repeated with multiparous individuals. The previously overdistended and obstetrically traumatized vaginas of multiparous women do not provide a satisfactory norm from which to measure significant vaginal distensibility other than on an individual basis. From the examination of individual women, however, the statement still can be supported that regardless of prior degree of vaginal expansion or increase in barrel length, the vagina will increase significantly in its measurable length and transcervical width under effective sexual stimulation.

During the excitement phase the vaginal walls also undergo a distinct color alteration. The purplish-red coloring of the normal steroid-stimulated vagina slowly changes to the darker, purplish hue of vasocongestion. This purple discoloration is patchy in character during the early stages of the excitement phase, but as plateau is achieved and pelvic vasocongestion becomes intense the entire vaginal barrel perceptibly darkens.

As excitement-phase reactions progress toward plateau, there is a flattening of the rugal pattern of the well-stimulated vaginal wall. This thinning or stretching of the vaginal mucosa is obviously a corollary of the involuntary expansion of the inner two-thirds of the vaginal barrel.

PLATEAU PHASE

Vaginal alterations in response to effective stimulation are not confined to the inner two-thirds of the vaginal barrel. There usually is a minimal distention of the outer third of the vagina during excitement phase. With attainment of plateau-phase levels of

sexual tension, a marked localized vasocongestive reaction develops
in this specific area of the vagina (Fig. 6-6). The entire outer
third of the vagina, including the bulbus vestibuli, becomes grossly
distended with venous blood. This vasocongestion is so marked that
the central lumen of the outer third of the vaginal barrel is reduced
by at least a third from the distention previously established during
the excitement phase. Although this localized vasocongestion develops as an involuntary response, it is a sure indication that
plateau-phase levels of sexual tension have been achieved. The
base of vasocongestion which encompasses the entire outer third
of the vagina, together with the engorged labia minora, provides
the anatomic foundation for the vagina's physiologic expression of
the orgasmic experience. This area of plateau-phase vasocongestion
has been termed the *orgasmic platform*.

A minimal further increase in width and depth of the vagina
(measured respectively at the transcervical depth and the fornices)
occurs during the plateau phase. This is a negligible degree of
clinical distention compared to the expansive response established

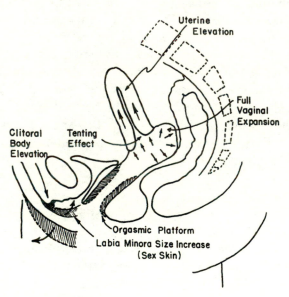

FIGURE 6-6

Female pelvis: plateau phase.

during the excitement phase. The degree of expansion is not standardized, so no attempt was made to measure it.

The production of vaginal lubrication reaches its full potential during the excitement phase. The production rate actually slows during the plateau phase, particularly if this level of sexual tension is experienced for an extended period of time.

ORGASMIC PHASE

The degree of expansion of the inner two-thirds of the vagina is not advanced beyond that attained during terminal stages of the excitement or early in the plateau phase of the sexual cycle. The basic reaction of the inner vaginal barrel is essentially expansive, however, rather than constrictive in character during the actual orgasmic experience.

The specific response of the vaginal barrel to the explosive physiologic entity of orgasm is confined to the orgasmic platform in the outer third of the vagina (Fig. 6-7). This localized area

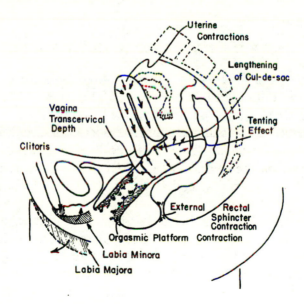

FIGURE 6-7
Female pelvis: orgasmic phase.

of bulbar vasoconcentration contracts strongly in a regularly recurring pattern during the orgasmic expression. The contractions
have onset at 0.8-second intervals and recur within a normal range
of a minimum of three to five, up to a maximum of 10 to 15 times
with each individual orgasmic experience. The intercontractile
intervals lengthen in duration after the first three to six contractions
of the orgasmic platform, and the measurable intensity of the
contractions progressively diminishes. The duration of the orgasmic
platform's recurring contractions and the degree of the contractile
excursions vary from woman to woman and within the same
individual from one orgasmic experience to the next. These recurrent contractions in the outer third of the vagina are the only
physiologic responses of the vaginal barrel that are confined entirely to the orgasmic phase of the sexual cycle.

At the highest tension levels ("status orgasmus"; see Chapter
9), the orgasmic platform may respond initially with a spastic
contraction lasting 2 to 4 seconds before the muscle spasm gives
way to the regularly recurrent 0.8-second contractions described
above.

RESOLUTION PHASE

With onset of the resolution phase, retrogressive changes develop
first in the outer third of the vagina (Fig. 6-8). The localized
vasocongestive concentration established during the plateau phase
which served as a platform for the spasmodic contractions of
orgasm is dispersed rapidly. As the result of this loss of localized
vasocongestion, the central lumen of the outer third of the vagina
actually increases in diameter in the early stages of the resolution
phase.

Slowly the expanded inner two-thirds of the vaginal barrel
shrinks back to the collapsed, unstimulated state. This reaction
is not one of a uniformly completed collapse of the vaginal walls.
It is rather an irregular, zonal type of relaxation of the lateral and
posterior walls. The anterior wall and the cervix of the anteriorly
positioned uterus descend toward the vaginal floor rapidly, quickly
resolving the tenting effect in the transcervical diameter of the

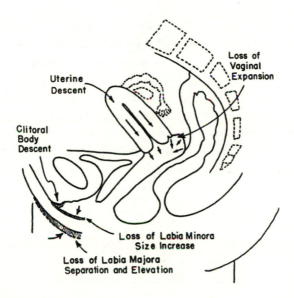

Uterine Descent

Loss of Vaginal Expansion

Clitoral Body Descent

Loss of Labia Minora Size Increase

Loss of Labia Majora Separation and Elevation

FIGURE 6-8

Female pelvis: resolution phase.

vagina. The return of the cervix to the vaginal axis from its elevated position in the pelvis and the coordinated collapse of the vaginal fornices frequently take at least three to four minutes to complete after orgasmic-phase release of sexual tensions.

The deep-purple color of the vaginal mucosa, developed during the plateau phase, returns to basic coloration in a slow retrogressive process which frequently requires as long as 10 to 15 minutes for completion. Normal rugal patterns, so typical of the hormonally well-stimulated vaginal barrel, become more apparent as the expansive reaction of the inner two-thirds of the vagina and the localized venous congestion of the outer third of the vagina retrogress. In rare instances the production of lubrication has been observed to continue into the resolution phase. This reaction continued beyond its normal stage of involution (late excitement or early plateau phase) suggests remaining or renewed sexual tension. Such individuals may be returned rapidly to orgasmic expression if stimulation is renewed.

The vaginal barrel's reaction to sexual tension has been con-

sidered in the restricted light of its role as the primary means
of the human female's sexual expression. The rapidity and intensity of the response mechanisms of lubrication production,
inner-barrel lengthening and expansion, transcervical tenting effect,
and vasocongestive development of the orgasmic platform always
parallel the degree of sexual tension experienced by the individual.
The vagina truly provides a direct physiologic reflection of female
psychosexual tensions, as it involuntarily prepares for and then accommodates the act of copulation.

2 . THE FUNCTIONAL ROLE IN REPRODUCTION

In performing its dual role, the vaginal barrel simultaneously
functions as a physical means of the human female's sexual expression and as an integral part of her conceptive mechanism. The
vagina's anatomic and physiologic reflection of woman's psychosexual tensions has been reported in Part 1 of this chapter. Its
only established role in conceptive physiology, that of a seminal
receptacle, is the area of present consideration. Even the mechanism
of this *functional* role has not been understood because vaginal
physiology has been, and still remains, an essentially unexplored
field. The possible *functioning* role of the vagina as an organ of reproduction rather than purely a functional seminal receptacle will
be approached in Part 3 of this chapter.

Following the usual pattern of medical response to investigative
challenge, many relatively obscure physiologic reactions of the female reproductive tract have been the target for more definitive investigative effort than have the more obvious reproductive functions
of the vagina. For instance, direct observation of human ovulation
has been accomplished [60]; tubal motility has been identified
[61, 274, 326]; and implantation of the fertilized ovum in the
wall of the uterus has been described in detail [127, 261]. To
date, the vagina's possible role as a functioning organ of reproduction and its role as a functional receptacle for seminal fluid have not
been considered in sufficient depth.

The anatomic angulation of the vagina assumes primary importance when the organ is considered purely as a seminal receptacle. The vaginal barrel normally angles about 10° to 15° below the horizontal pelvic line when a baseline is drawn from the vaginal outlet to the curve of the sacrum (Fig. 6-9). When a woman is in the supine coital position, the transcervical plane of the vagina is at a slightly lower level than the midplane of the fourchette of a nulliparous vaginal outlet. Therefore there is a natural gravitational tendency for a seminal pool to develop on the posterior wall of the inner half of the vaginal barrel. Obviously, there are numerous variations in vaginal anatomy and physiology which either improve, interfere with, or at times, even completely negate this normal pooling tendency.

The male partner must ejaculate if conception is to occur subsequent to coition (for exception see Chapter 14). Obviously, there is no requirement for orgasmic experience from the female partner. Women respond to coition with marked variation in sexual tension development. The male partner may ejaculate within the seminal receptacle while the female is in any of the four phases of the sexual response cycle. Therefore, the discussion of the anatomy of the

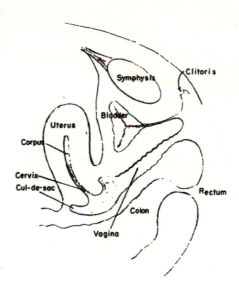

FIGURE 6-9
Female pelvis: vaginal-barrel angulation.

vagina as a functional seminal receptacle will be developed within the established framework of the sexual response cycle.

EXCITEMENT PHASE

Vaginal response to excitement-phase levels of sexual stimulation is confined to the inner two-thirds of the vaginal barrel. The primary physiologic reaction is an involuntary expansion and lengthening of the barrel. The vaginal wall lengthens in an anterior-posterior plane and expands in transverse diameter at the midcervical vaginal plane (see Fig. 6-3). With increasing tension, the anteriorly placed uterus is elevated into the false pelvis (see Chapter 8). Consequently the cervix is withdrawn from its sexually unstimulated positioning near the vaginal outlet, and elevated from the vaginal floor. The phenomenon of cervical elevation produces a bulbous or "tenting" expansion in the midvaginal plane. Thus, involuntary distention increases the circumferential diameter of the entire inner two-thirds of the vaginal barrel, with the greatest effect concentrated at the transcervical plane.

Further discussion of the basic physiology of vaginal accommodation to the mounting process is presented in Part 1 of Chapter 12.

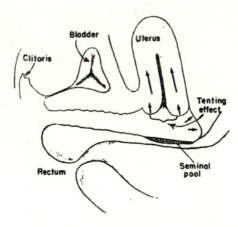

FIGURE 6-10

Nulliparous perineum: seminal-pool containment.

There are both anatomic variants and physiologic reactions that tend either to prevent or to encourage the escape of seminal fluid from the vaginal receptacle. An anatomic variant that aids seminal-fluid retention is the normally high, firm perineal support of the nulliparous woman. A high, tight vaginal outlet tends to trap seminal fluid in the vaginal depths with an actual damming effect (Fig. 6-10). The weight of the penis seldom is sufficient to overcome completely the constraining effort of the virginal perineum.

Conversely, an anatomic variant that encourages seminal-fluid wastage is observed when the virginal perineum has been altered by obstetric trauma (Fig. 6-11). A functional reproductive concern of the woman with an obstetrically traumatized perineum is the fact that a major portion of the total ejaculate may escape normal seminal pooling in the transcervical depths of the vaginal barrel. Without the damming effect of the nulliparous perineum there is a tendency toward immediate wastage of major portions of the seminal fluid immediately following penile withdrawal.

Many women with obstetrically lacerated or flattened vaginal outlets also develop an incipient or clinically defined rectocele. The existence of this anatomic variant frequently counterbalances the seminal-fluid-wastage tendency of the flattened vaginal outlet. The

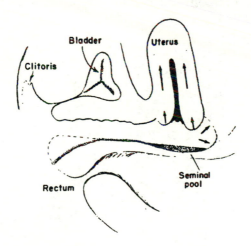

FIGURE 6-11

Parous outlet: seminal-fluid wastage.

rectocele's involuntary contribution toward seminal-fluid retention is improved by elevated sexual tensions. Excitement-phase expansion of the posterior vaginal wall is extended well beyond normal limits by the existence of separation or laceration of the levator sling (rectocele). The excessive expansion of the vaginal barrel at the transcervical depth made possible by the rectocele improves the vagina's retention of seminal fluid. In other words, the tendency of an anatomic variant to retain seminal fluid is improved by physiologic response to sexual stimulation.

PLATEAU PHASE

The major physiologic response of the vaginal barrel to plateau-phase levels of sexual tension is the creation of the orgasmic platform in the outer third of the vagina. This tension reaction is one of marked vasoconcentration (see Part 1 of this chapter). The localized vasocongestion creates a 50 percent constriction of the central lumen in the outer third of the vagina when compared to excitement-phase expansion in this area.

From an anatomic point of view, there is probably a greater chance of conception if the parous female achieves only plateau-phase levels of response during her sexual encounter, as opposed to

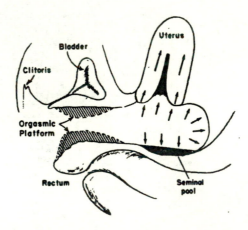

FIGURE 6-12

Plateau-phase orgasmic platform: seminal-pool containment.

enjoying an orgasmic experience. If only plateau-phase levels of tension are experienced, the orgasmic-platform vasocongestion will be dissipated at a much slower rate than that expected in a post-orgasmic sequence and consequently the physiologic aid in seminal-pool containment will be of longer duration (Fig. 6-12). If a woman does not obtain orgasmic release and must resolve target-organ vasocongestion from plateau-phase tension levels, the orgasmic platform may not be lost for 20 to 30 minutes. Obviously, the longer the external cervical os is exposed to an effectively contrived seminal pool the greater the possibility of conception.

The development of the orgasmic platform during advanced stages of female sexual response is an example of physiologic seminal-pool constraint. The orgasmic platform produces a stopperlike effect in the outer third of the vagina. This transitory constriction of the vaginal outlet helps retain seminal fluid within the transcervical depth of the vaginal barrel. Fortunately for reproductive interest, the orgasmic platform develops both in sexually responding nulliparous and multiparous females. Thus the tendency of the obstetrically traumatized woman to lose seminal fluid in the immediate postejaculatory period may be overcome or at least partially counterbalanced by the stopperlike effect of the orgasmic platform (Fig. 6-13).

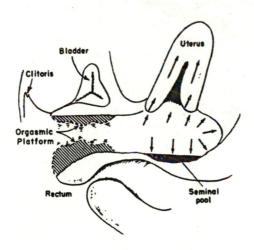

FIGURE 6-13
Parous outlet: orgasmic platform seminal-pool containment.

ORGASMIC PHASE

The physiologic constriction of the vaginal outlet which develops as an orgasmic platform does not remain indefinitely. If the sexually responding female experiences an orgasm, the localized vasocongestion in the outer third of the vaginal barrel is lost rapidly during her resolution phase.

With the anatomic variant of a retroflexed or retroverted uterus, the cervix may point in the vaginal axis or may be placed in direct apposition to the anterior vaginal wall. With the uterus in these posterior positions and the woman in supine coital positioning, the difficulty in establishing direct contact between the external cervical os and the seminal pool on the expanded posterior vaginal wall becomes obvious. If the uterus is fixed in severe third-degree retroversion, vaginal expansion in the midcervical plane essentially is unhampered. Only involuntary lengthening of the vaginal barrel is impeded.

Intercourse with the female partner in a knee-chest position will create a seminal pool on the anterior rather than the posterior vaginal wall. The cervix of the retroverted or retroflexed uterus has ready egress into such a seminal pool, if the recipient female remains in the knee-chest position after the male withdraws. A few minutes will be sufficient to allow effective contact between the external cervical os and the seminal pool created on the anterior vaginal wall by the knee-chest coital technique.

If only a plateau phase of sexual response is reached in the knee-chest position, the orgasmic platform created at this tension level will assist in constraining the subsequently developed seminal pool. If possible, orgasmic-phase response should be avoided for the female partner attempting to conceive in this position. Orgasmic experience would dispel the orgasmic platform from the vaginal outlet too rapidly to enable its constraining action on the seminal pool to be particularly effective.

RESOLUTION PHASE

Since we are considering the vagina as a functional seminal receptacle, a description of the anatomy of vaginal reaction subsequent to penile withdrawal is indicated. As the penis is withdrawn and the female enters the resolution phase of her sexual response cycle, there is a slow zonal relaxation of the expanded vaginal barrel to previously unstimulated positioning. The anterior vaginal walls and the cervix of the anteriorly placed uterus return to previous positioning more rapidly than the posterior or lateral walls of the vaginal barrel (Fig. 6-14). In this manner the cervix is quickly immersed in any normally constrained seminal pool in the transcervical depth of the vagina. That portion of the ejaculate not lost during penile withdrawal usually is well contained within the transcervical depths of the nulliparous vagina for a matter of hours, unless the individual female assumes other than a supine position. There is an earlier loss of seminal fluid from the parous woman with an obstetrically traumatized perineum. Here seminal fluid is lost as soon as the orgasmic platform is resolved. Excessive pelvic activity such as elevation to a sitting or standing position or even changing position from side to side in bed will occasion further loss of seminal fluid.

Two other variants, based on male postejaculatory behavior, tend to dispel the local concentration of the intravaginal seminal pool and should be noted briefly. If the sexual unit continues in quiescent coital connection after the male has completed his ejaculatory

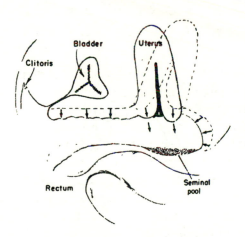

FIGURE 6-14

Cervical immersion in seminal pool. Resolution phase.

effort, the weight of the penis acts to flatten the posterior wall in the outer third of the vagina. This will result in escape of seminal fluid from the natural pooling in the transcervical depth of the vagina. For the parous woman, penile weight also will tend to overcome the constrictive effect of an orgasmic-platform residual at the vaginal outlet.

Some males attempt to continue active coital thrusting after ejaculation and before loss of full erection. This situation usually develops when the male is trying to accomplish orgasmic release for his tension-distressed partner. Seminal-fluid loss is excessive in this instance. There is no opportunity for seminal-pool formation, and during active penile stroking each withdrawal movement encourages further fluid escape.

Thus the efficiency of the vagina's functional role as a seminal receptacle is influenced primarily by the anatomic and physiologic factors of vaginal angulation, sex tension-induced transcervical expansion, parity, obstetric trauma, orgasmic-platform constriction of the vaginal outlet, and coital positioning. A secondary influence on the vagina's functional role is exerted by the male partner through improper coital technique.

3 . THE FUNCTIONING ROLE
IN REPRODUCTION

The possibility that the vagina has a *functioning* role as an organ of reproduction rather than purely a *functional* role as a seminal receptacle first was suggested by Sims seventy-five years ago [297]. Unfortunately, minimal investigative interest has been directed toward the analysis of multiple factors in the intravaginal environment which, theoretically at least, either would be supportive of or antagonistic toward spermatozoal migration. Spermatozoa must be able to exist in the vaginal environment and migrate through the vaginal barrel into the ovulatory cervical mucus if postcoital conception is to occur.

Brief mention should be made of some of those who have

devoted serious consideration to the possible role of the vagina as a functioning organ of reproduction. In 1913 Huhner popularized Sims' original concept that microscopic examination of postcoital cervical mucus was of primary importance in any fertility examination. He described this postcoital examination of cervical mucus as the *Huhner test* and established standards for the test [121, 123, 124]. Subsequently, Séquy and colleagues [286, 287] refocused medical attention on Sims' original suggestion of the importance of timing postcoital examination of cervical secretions to coincide with what we now know to be ovulatory receptivity to the spermatozoa. Unfortunately, in 1930 Moench's [226, 227] tenet that there was little of consequence within the vagina to influence longevity or motility of spermatozoa gained wide credence within the medical profession. As a result of his influence, the vagina was considered for many years to be only a functional, not a functioning, part of the female reproductive system. It remained for Rakoff [253, 254] and Lang [164, 165] to provide the present impetus for investigation of vaginal conceptive physiology. Their findings serve as a foundation for any investigative departure into the functioning role of the vagina in reproductive physiology.

Vaginal environment is subject to influences from many sources. The fact that the vaginal mucosa is a permeable two-way membrane [179, 203] emphasizes the difficulty in establishing and maintaining a stable environment. The potential factor of contamination of the vaginal barrel by infectious agents [34, 35, 218, 264, 283] must always be considered. (These fluctuating influences on vaginal environment are mentioned only in passing, as space does not permit a more definitive consideration. Only concerns of possible sensitivity reactions and of acid-base response to seminal-fluid influence will be considered at this time. Recent publications [256, 272, 298] point up the functioning role of intravaginal environment in reproductive physiology.)

Since Zweifel [333] first described the normally acid state of vaginal discharge, the source of vaginal acidity and concerns of in vitro pH influence upon seminal elements have been of interest to numerous investigators [96, 236, 237, 243, 288, 289, 321]. Generally it has been accepted that normal vaginal acidity ranges from a pH of 4.0 to 5.0. Significant pH elevation has been related to cyclic luteal influence on the vaginal mucosa. This information was made available by Rakoff [254], using a modification of the Trussell and MacDougal technique [317] for estimating acidity of the vaginal environment.

Cruickshank and Sharman [47] emphasized the many factors which influence vaginal acidity. The pH of the vagina is a secondary reflection of the effective production of ovarian hormones. Vaginal pH is developed and maintained by the sloughing of mature cells of the superficial layers of the vaginal mucosa. These cells under estrogenic influence contain a high concentration of stored glycogen which is, in turn, metabolized to lactic acid within the vaginal canal.

After Muschat and Randall [236, 237] pointed out hydrogen-ion influence on motility of human spermatozoa, a number of experiments have clarified optimum conditions for sperm motility in an acid media such as that present in the normal vaginal environment. Voge [321] first reported that a pH of 4.9 to 6.4 definitely arrests sperm motility. He also pointed out that once spermatozoa are exposed to a pH of 3.5 or below, revival of motility by neutralization of the acid environment was not possible. It remained for Shedlovsky [288, 289] to provide the most definitive work in this field, when he completely immobilized spermatozoa in less than one-half minute in a pH greater than 12.2 and in a pH of less than 3.0.

Huhner adamantly defended the thesis that neither the viability of spermatozoa in secretions of the female genital tract nor their actual value for purposes of impregnation can be determined by any chemical or physical test of the seminal specimen itself [122]. He suggested the possibility that the female genital secretions might have even more power to preserve the vitality of spermatozoa than the seminal fluid proper [123, 124]. His suggestion that there might be a functioning role for the vagina in reproductive physiology was amplification of Sims' pioneer thinking on this subject. Suffice it to say that there are many influences on sperm motility and longevity which develop within, and are a natural part of, vaginal environment. To date, most of these influences not only are unexplored but are also without even hypothetical definition.

The unfortunate part of all of these early studies is the fact they were in vitro determinations, and no in vivo intravaginal correlation was attempted. Laboratory climate and intravaginal environment are completely separate entities. There is a specific clinical application of information gathered by in vitro investigation of the seminal fluid and its contents. For instance, the in vitro studies [289] describing human spermatozoal inability to be reactivated after an exposure to a pH of 3.5 have been proved to have no significant clinical application by in vivo observations. We now know that a controlled vaginal environment normally will maintain a pH ranging from 3.5 to 4.0. Controlled vaginal environment arbitrarily has been defined as a vaginal barrel without pathogenic bacteria or fungi, and with no exposure to the neutralizing effect of seminal fluid or menstrual flow for a 48-hour period.

Another factor in vaginal environment which may have influence upon spermatozoal motility or longevity is vaginal lubrication. This is the substance that develops with sexual tension as the first evidence of the human female's excitement-phase response to sexual stimulation (see Part 1 of this chapter). It appears on the walls of the vagina within 10 to 30 seconds after reception of either somatogenic or psychogenic sexual stimulation. The longer foreplay is extended, and the longer coition is continued without ejaculation, the greater the total production of the vaginal lubrication. The influence of lubricating material on vaginal acidity is related directly to the amount produced and the duration of its production [202, 206].

With vaginal acidity established in the 3.5 to 4.0 pH range, approximately 30 minutes of lubrication production must occur before influence on such a low pH is recordable. Elevation of a baseline pH (3.5–4.0 in the unstimulated vagina) into the range of 4.25–4.5 has been recorded repeatedly after relatively extended periods of sexual activity. The responding female may experience multiple orgasms or only achieve plateau-phase levels of sexual tension during extended manipulative and coital episodes. The one constant factor influencing the production of vaginal lubrication is long-maintained stimulative activity, regardless of type employed, whether or not orgasmic level of sexual response is obtained. This statement presumes, of course, that the female at least reaches an excitement-phase level of sexual response.

Since the pH of lubricating material obviously has a higher range than that encountered in a controlled vaginal environment, it may be that this lubricating material also plays a preordained role in adjusting vaginal acidity to provide the most effective environment possible for spermatozoal migration. It will be recalled that there is a marked buffering power in seminal-fluid content. How much the vaginal lubrication, through its influence on baseline vaginal acidity, aids the inherent buffering power of the seminal fluid to protect spermatic motility by elevating vaginal pH has not been determined.

Laboratory standards for fertile and infertile males have been established in accurate detail by MacLeod and his co-workers [184–190]. His exhaustive efforts have provided the medical profession with a baseline for effective in vitro evaluation of male fertility. However, the clinician investigating an infertile family unit may make a serious mistake by defining male fertility only on the basis of an in vitro evaluation of a manually produced seminal specimen. The in vitro evaluation of the male partner is the step of primary importance and is of tremendous value, but it is not the only step to be taken.

A second important step (which should be part of every evaluation of an infertile family unit) is definition of spermatozoal viability and motility after intravaginal deposition of the ejaculate. The Sims-Huhner type of marital-unit evaluation (microscopic examination of postcoital ovulatory cervical mucus) should be just as constant a part of the infertility work-up as the in vitro seminal specimen examination [34, 211].

The immediate effect on vaginal environment created by the ejaculatory deposition of seminal fluid has not been fully appreciated. In an attempt to illustrate this effect, measurements of intravaginal pH were conducted after the technique popularized by Rakoff et al. [254]. Due to the immediacy of the buffering effect of seminal fluid, glass measurement electrodes were standardized at a pH of 7.0 routinely. Therefore slightly higher readings for baseline vaginal acidity were recorded during these experiments than would represent the true value if the electrodes were standardized with a pH of 1.0. Despite this experimental inadequacy all measurements of the pH of controlled vaginal acidity fell in the 3.5–3.9 range.

In order to prevent possible intravaginal contamination a period of vaginal control (3–7 days) was established for the family units cooperating with the investigation. The vaginas all returned negative cultures for bacterial or fungal pathogens within a 48-hour period before the experimental observations. Details of these experiments have been reported [206, 208]. To avoid repetition, the results reported will be restricted to those obtained from only one of the experimental sessions.

As recorded in Figure 6-15, there was a total stimulation time of 17 minutes (vaginal lubrication production) from the onset of foreplay until the male's ejaculatory experience. The first measurement of vaginal acidity and examination of vaginal content was initiated 9 seconds after intravaginal deposition of seminal fluid. Readings of vaginal acidity were recorded at 10-minute intervals for 2 hours, and then every hour for 10 hours. A concluding

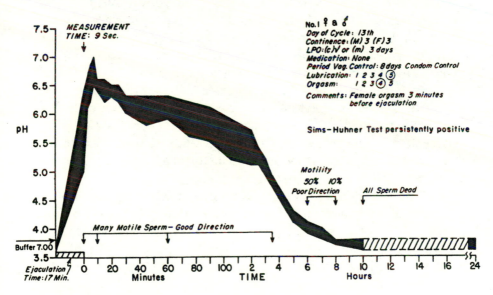

FIGURE 6-15

Vaginal environment and seminal-fluid content (in vivo determinations).

estimate of postejaculatory vaginal acidity was recorded 24 hours after coition.

Neutralization of total vaginal acidity occurs immediately after ejaculation. Although the major part of the seminal fluid usually is delivered to the transcervical depth of the vagina, penile withdrawal serves to spread portions of the ejaculate throughout the vaginal barrel. A sufficient amount of seminal fluid escapes the transcervical intravaginal pool to neutralize total vaginal acidity. The buffering effect of this particular ejaculate was sufficient to provide a residual influence on total vaginal acidity that lasted for 6 to 7 hours. For other family units the active neutralizing action of the ejaculate has been recorded for as long as 16 hours.

In order to explore this aspect of the vagina's functioning role in reproductive physiology at the most plausible time, measurements were taken on the thirteenth day of a regularly recurring 28-day menstrual cycle (see Fig. 6-15). This fertile study subject (as well as other subjects) was determined to be in an immediate preovulatory phase as defined by vaginal cytology [294]. There was a three-day continence period for the family unit. The last prior female orgasm had occurred three days previously, during coition that was condom-controlled to avoid vaginal contamination. An excessive amount of vaginal lubrication developed during the experimental period and was designated as Grade 5 by the investigators (the highest possible grade). The study subject personally graded her orgasm at a level of 4, which is the highest grading assigned to a single orgasmic experience (Grade 5 represents multiple orgasm). The female study subject achieved orgasm approximately three minutes before her male partner ejaculated.

Further evidence suggesting a normal functioning role for the vagina of this study subject was acquired during the three months prior to the experimental period by evaluating spermatozoal motility in the vaginal environment. The Sims-Huhner tests of postcoital ovulatory cervical mucus were consistently positive for this family unit.

At the time of initial pH measurement a specimen of vaginal content was obtained (see Fig. 6-15). This specimen of seminal fluid microscopically showed many spermatozoa motile with good direction. The postcoital vaginal content was examined in similar fashion every 15 seconds for the first postcoital minute. Thereafter, specimens were obtained every minute for 5 minutes, and again at the 10-minute interval. The vaginal content routinely was examined at 1, 3, 6, 8, 10, and, finally, 24 hours postcoition. All specimens were examined microscopically to evaluate spermatozoal longevity and directional motility within the functioning environment of the postcoital vagina. Spermatozoa continued motile with good direction until the 10-hour examination. At this time and thereafter, all spermatozoa were nonmotile.

In substance, Figure 6-15 records the response of a functioning vagina to the coital deposition of a fertile seminal specimen. The chart is representative of 53 observations of fertile family units conducted under similar conditions, and establishes a baseline for comparison with infertile family units. Fertile family units have been evaluated at various stages of the female partner's menstrual cycle. No significant variation in hydrogen-ion influence on intravaginal spermatozoal longevity or directional motility has been recorded from the immediate postmenstrual to late premenstrual phases of the menstrual cycle. It also makes no difference in intravaginal spermatozoal motility or longevity whether seminal fluid is deposited by coition or by insemination techniques.

A routine in vitro evaluation of all male partners was conducted from 3 to 5 days after the experimental sessions with the fertile family units. In every instance this examination was carried out on the first seminal specimen produced, after that measured in the family-unit investigation. In this way, the male partner's levels of fertility as described by these in vitro studies could be compared with the family unit's in vivo results.

In addition to the routine in vitro examinations (seminal-fluid volume and viscosity; sperm count per cubic centimeter, directional motility, longevity, and morphology), the seminal specimen was subjected to a four-day check of hydrogen-ion concentration (Fig. 6-16). This technique was used to demonstrate the stable, almost incredible buffering power inherent in seminal-fluid content. The manually produced specimen was maintained at 37°C. over a water bath without carbon dioxide contact. Dicrysticin was added to the seminal fluid to control possible bacterial contamination. Al-

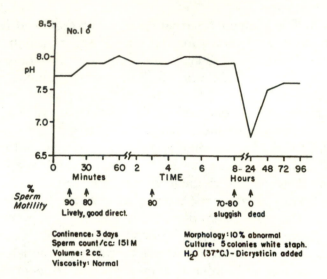

FIGURE 6-16

Seminal-fluid content (in vitro determinations).

though the specimen was obtained under the most sterile conditions possible, routine anaerobic and aerobic cultures were taken as
a means of experimental control. In this instance, bacteriologic
examination described five colonies of *Staphylococcus albus*, which
for purposes of this study was considered to be nonpathogenic in
character.

Routine evaluation of this fertile semen specimen showed a
volume of 2 cc. and essentially normal viscosity. The inadequate
volume of the ejaculate was counterbalanced by the sperm count
which was reported as 151 million per cc. The abnormal spermatozoal morphology was estimated at 10 percent. Observation at eight
hours demonstrated spermatozoa still active; however, directional
motility was reduced significantly. The only basic difference between the in vitro and in vivo evaluation of this male's fertile seminal fluid was the fact that the spermatozoa were shown to live
longer under the somewhat controlled in vitro conditions. There
is, however, real doubt of the fertilizing ability of spermatozoa after
even six to eight hours of controlled in vitro containment.

Abnormal variants of conceptive physiology are being identified as investigative interest is directed toward the possible functioning role of the vagina as a reproductive organ. The recently established "lethal factor" exists in the vaginal environment and immobilizes apparently healthy spermatozoa [208, 211]. This example of abnormal physiology, in its acute stage, produces an infertile state for any family unit contending with the problem. Influence of the "lethal factor" on vaginal conceptive physiology is shown in Figure 6-17, which records information returned from the evaluation of an infertile family unit.

The Sims-Huhner test was negative on three separate occasions. The experiment reflected in Figure 6-17 was conducted on the fourteenth day of a regularly recurring 28-day menstrual cycle. Vaginal cytology placed the female partner in an immediate preovulatory phase. There had been a seven-day continence period for both partners. The period of vaginal control also was one week.

The last prior orgasm had been experienced seven days previously. The female partner achieved only plateau-phase tension levels dur-

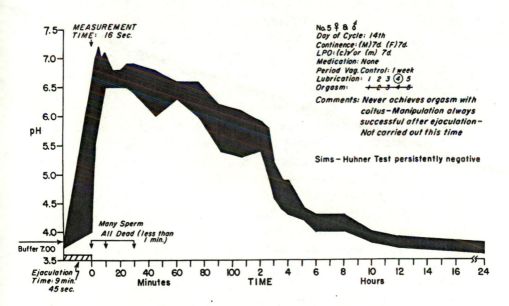

FIGURE 6-17

The "lethal factor": vaginal environment and seminal-fluid content (in vivo determination).

ing the experimental session. This individual has never achieved the orgasmic phase of sexual response with coitus. However, manual manipulation has always provided orgasmic release and is routinely employed by this family unit after the male ejaculatory experience. In order to provide for immediate measurement of post-coital vaginal content, this technique of female sexual tension release was avoided. A Grade 4 degree of vaginal lubrication developed during the experimental session.

As indicated in Figure 6-17, seminal-fluid influence upon total vaginal acidity was measurable 16 seconds after the male partner's ejaculation. The buffering power of this particular seminal specimen was exceptional. There was an immediate rise from the baseline vaginal pH levels of 3.7 to 3.9 into the range of 6.5 to 7.2 pH. The marked buffering effect of this particular seminal fluid against the total vaginal acidity maintained the intravaginal environment at 6.5 pH level approximately 80 minutes after ejaculation. The pH of the vaginal environment then fell slowly but steadily toward a level of 4.0, which was recorded 9 hours after the male's ejaculatory experience. At the 24-hour check, vaginal acidity was just below the 3.7–3.9 pH levels recorded at the onset of the experiment.

Although there were many spermatozoa in specimens of vaginal content taken subsequent to ejaculation, all spermatozoa were observed to be completely immobile. Specimens were returned for examination following the time schedule described earlier in the chapter in discussion of normal vaginal physiology. The influence of the "lethal factor" was established even in the sample of vaginal content removed just 16 seconds after intravaginal seminal deposition. Every additional sample of vaginal content demonstrated immobile spermatozoa.

The in vitro evaluation of the male partner followed the family-unit experimental session by four-and-a-half days (Fig. 6-18). The manually produced seminal specimen showed a normal count of 95 million per cc., a total volume of 5 cc., and a moderately increased viscosity of the seminal fluid. The incidence of abnormal spermatozoal forms was somewhat high (recorded in the 15 percent range).

Initial examination of the seminal specimen showed 70 to 80 percent directional motility. At the three-hour check, there continued

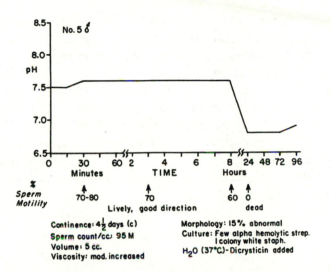

8.5
No. 5 ♂
8.0
pH
7.5
7.0
6.5

0   30   60 2   4   6   8 24 48 72 96
     Minutes   TIME   Hours

%
Sperm
Motility
         70-80   70   60 0
            Lively, good direction   dead

Continence: 4½ days (c)   Morphology: 15% abnormal
Sperm count/cc: 95 M   Culture: Few alpha hemolytic strep.
Volume: 5 cc.   I colony white staph.
Viscosity: mod. increased H_2O (37°C.) - Dicrysticin added

FIGURE 6-18

The "lethal factor": seminal-fluid content (in vitro determinations).

to be 70 percent directional motility, and at eight hours 60 percent directional motility. The 24-hour check showed all spermatozoa immobilized. The specimen was produced under as aseptic conditions as possible. However, anaerobic and aerobic cultures of the ejaculate demonstrated a few colonies of alpha-hemolytic streptococci and one colony of *Staph. albus*. The bacteriology suggested contamination rather than incidence of pathology.

Examination of the pH of the seminal specimen showed a drop just below the 7.0 level at the 24-hour observation. The rest of the four-day observation period demonstrated a slow rise in recorded pH for this seminal specimen. The specimen was maintained at 37°C. over a water bath without CO_2 contact. Dicrysticin was added to control possible bacterial contamination. Again the buffering power of the seminal-fluid content was demonstrated in the manner established and described above. This male would have to be judged (by laboratory standards) as essentially a fertile individual.

In an evaluation of the family unit, it is obvious that their particular conceptive concern is one of absolute spermatozoal immobility which develops immediately subsequent to ejaculatory deposition of the seminal-fluid content in the vaginal environment.

This family unit represents definition of the instant lethal effect of an intravaginal environment upon apparently normal spermatozoa. Thus, there is proof positive of an obvious functioning role for the vagina in conceptive and contraceptive physiology. There will be a great deal more evidence of the vagina's functioning role in conception when more definitive research has been designed, conducted, and reported. To date little has been done, and little is known of the vagina's conceptive functioning role despite the obvious availability of the organ to investigative procedure.

7

THE ARTIFICIAL VAGINA
ANATOMY AND PHYSIOLOGY

Anatomic and physiologic response patterns have been described for the normal human vagina in its dual role as the functional means of the human female's sexual expression and as a functioning part of her reproductive mechanism. Of comparative interest, from a functional point of view only, are the anatomic and physiologic response patterns of the artificial vagina.

The material to be presented has been accumulated from follow-up evaluation of seven women, ages 19 to 34 years, who were treated for vaginal agenesis (congenital absence of the vagina). Five of these anomalies were resolved by surgery; two responded to mechanical methods. There is no necessity for a description of either the surgical or the mechanical techniques used in these cases, since they have been described in detail by numerous authors in previous publications [4, 31, 45, 72, 77, 78, 92, 183, 223, 281, 327, 328]. The method of creating an artificial vaginal barrel is incidental, since the functional reaction patterns of artificial vaginas are identical regardless of how they are constituted.

Psychosexual histories obtained from individual subjects will be omitted. While the varied backgrounds which produced seven vastly different modes of psychosexual orientation are of interest, they are of no statistical value. Psychosexual concerns are not relative to the physiology being presented, and an adequate overall interpretation of the sexually functional role of these vaginas can be established on the basis of essentially identical physiologic reaction patterns.

Since the anatomic and physiologic response patterns of the artificial vagina are to be considered in comparison with those of the normally constituted vagina, the four phases of the human female's sexual response cycle will serve as a descriptive framework.

EXCITEMENT PHASE

Following the pattern of normal vaginal reaction, the first anatomic evidence of an excitement-phase response to sexual stimulation appears in the form of a mucoid material on the walls of the artificial vagina. This lubrication appears in droplet formation in a matter of 30 to 40 seconds after the onset of any form of somatogenic or psychogenic sexual stimulation. Production of the lubricating material in clinically sufficient amounts usually takes longer than in the normal vagina. Long-maintained sexual tension levels or extensive foreplay usually is necessary for achievement of an effectively lubricated vaginal barrel prior to coition. This slowed rate of initial response should not be confused with totality of performance for the artificially constituted vagina.

With the first few penile strokes the production of lubrication is increased and coalescence of the material is accomplished rapidly. The accelerated production of lubrication that develops in direct response to active coition is a reactive tendency characteristic of the artificial vagina. It should be stated, however, that some surgically or mechanically created artificial vaginas are capable of lubricating as well and as rapidly as any normally constituted vaginal barrel. Two of the seven women under evaluation have been observed to lubricate extensively and rapidly in response to sexual stimulation—in fact, more effectively than many women with normally constituted vaginas. In any event, the phenomenon of lubrication makes an artificially constituted vagina a highly functional seminal receptacle readily available for normal coition.

Speculation as to the actual physiologic source of vaginal lubrication continues to be of some gynecologic interest. Just as with the normal vagina, sexual tensions produce marked dilatation of the venous plexus concentration which encircles the entire artificial vaginal barrel. The mucosa of the artificial vagina has been established as a functioning two-way membrane, as is normal vaginal mucosa. At present there is no clinical contraindication to the presumption that spontaneous lubrication of the artificial vaginal barrel comes from the same source and, in all probability, has the same chemical constituents as that of the normal vaginal barrel [202].

It should be stated parenthetically that in five of the clinically resolved cases of vaginal agenesis lubrication has been observed to be more effective during the luteal than the estrogenic phase of recurring hormonal cycles. Since only seven cases of vaginal agenesis comprise the total number that have been evaluated to date, there is no statistical significance to this observation.

As the excitement-phase continues, secondary anatomic reactions to elevated sexual tensions develop within the artificial vagina just as they do within the normal vagina. As previously described for the normal vagina (see Part 1 of Chapter 6), there is a lengthening and distention of the inner two-thirds of the vaginal barrel (Fig. 7-1). The area in question alternately expands spasmodically, and then relaxes in a slow, tensionless manner essentially similar to that of the normally constituted vagina.

In addition to lengthening of the vaginal barrel, there is an obvious diameter increase in the inner two-thirds of the artificially

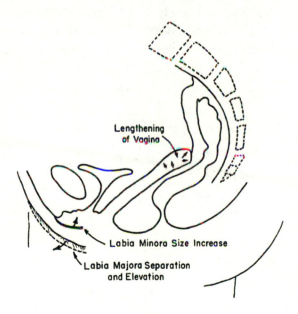

FIGURE 7-1
Artificial vagina: excitement phase.

created vagina. However, elasticity of mucosal movement is reduced significantly when compared to that of the normal vagina. Both expansive and lengthening abilities develop more slowly without the fluidity of movement and degree of excursion that normal vaginal walls possess. Just as the normal vagina is a clinically potential rather than an actual space, so the anterior and posterior walls of the artificial vagina are collapsed together except when under the influence of a significant degree of sexual tension.

The involuntary lengthening and distention of the artificial vaginal barrel has been measured in a manner similar to the technique used for the normally constituted vagina. Due to the essential uniformity in physiologic response of the artificial vaginas, it is possible to report one study subject (Subject A) as representative of the general response patterns for other individuals evaluated.

The vaginal length and midvaginal diameter of Subject A have been measured repeatedly. These measurements have been taken routinely from the fourchette to the depth of the vaginal barrel and at the point of greatest expansion of the central diameter, which usually develops approximately 2 cm. from the distal end of the vagina (Fig. 7-2). When Subject A is stimulated sexually to advanced excitement-phase levels, the expanded central diameter

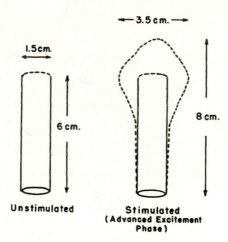

FIGURE 7-2

Artificial vagina: baseline measurement, Subject A.

measures approximately 3.5 cm. as opposed to the unstimulated central diameter measurements of 1–1.5 cm. The artificial vaginal barrel increases in length to an average of 8 cm. from an unstimulated length of 6 cm.

Subject A also demonstrates the artificial vagina's ability to effect further vaginal-wall expansion under the influence of continued or increased sexual stimulation (Fig. 7-3). With the anterior and posterior speculum blades fixed at a constant 2 cm. separation, the initial lateral wall (central diameter) expansion developed by speculum placement was measured at 2.5–3 cm., as opposed to the 1–1.5 cm. of the unstimulated state. The length of Subject A's artificial vaginal barrel was increased from 6 to 7.5 cm. by the placement of an indwelling speculum before conscious sexual stimulation was initiated.

When the subject was stimulated sexually to an advanced excitement phase with an indwelling speculum in the vaginal barrel, the point of greatest lateral wall expansion (central diameter) measured 3.5–4 cm. in width, while the length of the vaginal barrel was increased to approximately 9 cm.

As a general rule, the more the artificial vagina is dilated, the greater the potential for further dilatation. While the degree of dis-

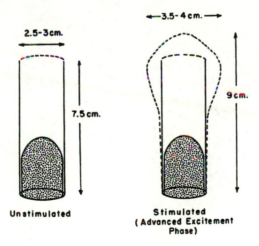

FIGURE 7-3

Artificial vagina: speculum-dilated, Subject A.

tensive ability is not that of a normal vagina, certainly this inherent functional response to female sexual tension remains a factor of clinical significance. The artificial vagina under effective sexual stimulation can accommodate any size of penis if there has been successful surgical or mechanical development.

PLATEAU PHASE

When plateau-phase levels of sexual response are achieved, localized vaginal vasocongestion is apparent. This reaction develops with such severity that there is approximately a 50 percent occlusion of the central lumen of the vaginal barrel. Just as an orgasmic platform develops in the outer third of the normal vaginal barrel, so an orgasmic platform constantly develops in the outer third of the artificial vaginal barrel (Fig. 7-4).

The production of vaginal lubrication increases markedly during the plateau phase, as opposed to the somewhat slowed production

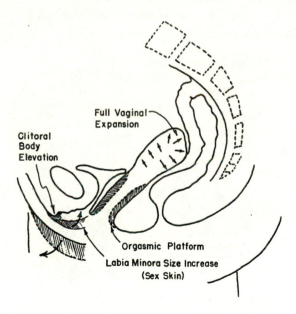

FIGURE 7-4

Artificial vagina: plateau phase.

rate of the normal vagina at this tension level. The highest rate of lubrication production in the normal vagina is confined to the excitement phase of the sexual response cycle. There is relatively little increase in central diameter or length of the artificial vaginal barrel during the plateau phase of sexual response.

The labia minora of women with an artificial vagina turn the sex-skin color of bright red, just as do the labia minora of the normal nulliparous women, when an advanced plateau phase of sexual excitement is attained (see Chapter 4). When this sign of impending orgasm occurs (presuming that effective sexual stimulation is continued), orgasm is sure to follow in women with an artificial vagina, just as it does in women with the normally constituted vaginal barrel.

ORGASMIC PHASE

The characteristic physiologic expression of orgasm in both artificial and normal vaginas is the onset of regularly recurring contractions of the orgasmic platform (Fig. 7-5). This platform, created by plateau-phase vasocongestion, contracts strongly and in a regularly recurrent pattern during the orgasmic phase of the female sexual cycle. The contractions have a frequency rate, at onset, of approximately 0.8 second. The intercontractile intervals slow as the orgasmic experience progresses. Orgasmic-platform contractions of artificial vaginas have been measured by gross pressure-gradient techniques. They recur with slowly increasing intercontractile intervals on an average of from 5 to 10 times. The physiologic contractions of the orgasmic platform increase in number in direct parallel to the severity of the experience.

The orgasmic response of the individual with an artificial vagina includes involuntary contractions of the entire perineal body. Not only the outer third of the vagina but also the external rectal sphincter and the lower abdominal musculature contract in orgasmic-phase expression. The superficial and deep transverse perineal, the bulbospongiosus, the levator ani, and the lower portions of the rectus abdominis muscles are the muscles of primary response. Once the artificially created vagina becomes available to

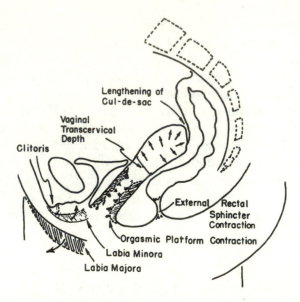

FIGURE 7-5
Artificial vagina: orgasmic phase.

heterosexual function, both voluntary and involuntary muscular control develops with normal physiologic responsiveness.

The one completely characteristic reaction of orgasmic-phase response in artificial vaginas, never observed in normal vaginas, is a marked color change which occurs throughout the entire vaginal barrel. Vivid discoloration of the barrel appears with sudden onset during the actual orgasmic experience. This phenomenon of orgasmic color change has been recorded by cinematography.

The mucosa of the artificial vagina varies in color from gray to purplish-red in a sexually unstimulated state. During an orgasmic experience, the mucosal color changes to a bright red. This flash of red is startling both in suddenness of onset and in vividness of coloration. The more intense the color change, the more intense is the orgasmic experience. This orgasmic color reaction has never been identified in normal vaginas during hundreds of directly observed sexual response cycles. The trigger mechanism for this marked vasocongestive reaction to orgasmic experience has not been defined.

RESOLUTION PHASE

Following the involutional pattern established for the normal vagina, the retrogressive changes of the resolution phase in the artificial vagina occur in reverse order of their original development (see Part 1 of Chapter 6). The first resolution-phase responses are loss of the localized vasoconcentration in the outer third of the vagina (orgasmic platform) (Fig. 7-6) and involution of the sex-skin discoloration of the labia minora. Actually, the sex-skin discoloration disappears more rapidly than does the orgasmic platform in women with an artificial vagina.

The walls of the artificial vaginal barrel slowly shrink back to the unstimulated baseline measurements. Just as the expansion reaction of the artificial vagina is delayed in time sequence, when compared to that of the normal vagina, so the return of artificial vaginal walls to an unstimulated collapsed positioning occurs at a much slower rate than that recorded for the normal vagina.

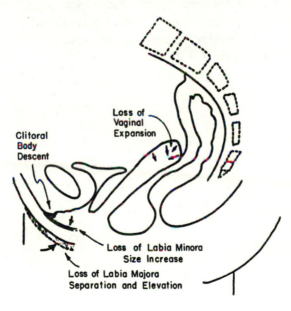

FIGURE 7-6
Artificial vagina: resolution phase.

The orgasmic color change is slow to resolve. Residual discoloration frequently can be demonstrated for as long as 10 to 15 minutes after women with an artificial vagina have experienced physiologic release from sexual tensions.

If the artificial vaginal barrel has been created successfully either by surgical or mechanical means, the social catastrophe of vaginal agenesis may be resolved. The artificial vaginal barrel, once successfully constituted, is a thoroughly effective coital mechanism. Generally, reactions to effective sexual stimulation are somewhat delayed in rapidity and intensity, when compared to those of the normal vaginal barrel. The successfully created artificial vagina rapidly and effectively assumes its proper functional role as the primary physical means of the human female's sexual expression.

The psychosexual problems that develop subsequent to the creation of an artificial vagina have not been dealt with in this chapter. Prior publication has discussed at some length the psychosocial concerns of three women with a successfully constituted artificial vagina [209]. Since there were only seven study subjects available, statistical support was insufficient for a generalized psychosexual discussion. A consideration of the psychosexual and social problems inherent in the clinical condition of vaginal agenesis will be presented at a later date.

8

THE UTERUS
PHYSIOLOGIC AND CLINICAL CONSIDERATIONS

Despite the milestone of the Reynolds contribution [261], relatively little is known of human uterine physiology in those areas not related directly to states of pregnancy or hormonal influence on the myometrium. Due to the technical difficulties inherent in working objectively with the human uterus, information available usually has been developed in a fragmentary manner. For example, the cyclic influence of ovarian hormones on the secretory activity of the glands in the uterine endometrium is well established [149], yet the biochemical constitution of the material secreted by these glands remains an unresolved problem. Moreover, areas of potential influence for this material either locally within the uterine cavity or distally in some target organ such as an ovary have not been established.

A similar picture reflects the levels of knowledge of uterine-body response to sexual stimulation. Finite details of uterine-body reaction to sexual stimuli will not be established until the problem is attacked by a more definitive program of physiologic investigation than has been possible. Sufficient inroads have been made by the combined techniques of basic physiologic recording and direct clinical observation to establish certain truths in the culturally resistant area of uterine response to sexual stimulation. The uterus has been observed to respond to sexual stimulation as a composite organ. In addition, individual reaction patterns for the corpus and the cervix have been recorded and observed. The four phases of the cycle of sexual response will be employed to facilitate a discussion of reactions of the organ as a whole and of those of its composite parts.

The first phase of uterine response to sexual stimulation has been established by considering the organ as a whole rather than by focusing on either of its component parts, the corpus or the cervix. This total-uterine-body reaction develops at excitement-phase levels of sexual tension and is related directly to uterine positioning in the female pelvis. The following description of uterine-body reaction to sexual stimulation presumes an anteriorly placed or midpositioned uterus and does not apply to the organ in retroverted or retroflexed pelvic positioning.

UTERINE-ELEVATION REACTION

As excitement-phase levels of sexual tension progress toward the plateau, the entire uterus elevates from the true into the false pelvis. Under direct observation, the cervix is removed from its normal resting position in direct contact with the posterior vaginal floor, not far from the vaginal introitus. The cervix slowly retracts from its resting position in a posterior and superior plane as the vaginal walls expand under the influence of sex tension increment (Fig. 8-1).

As the entire uterus elevates into the false pelvis, the cervix, together with the involuntarily expanding anterior and lateral vaginal walls, creates a tenting effect in the transcervical depth of the vaginal barrel (see Part 1 of Chapter 6). The elliptical expansion in the midvaginal plane resultant from cervical elevation and anterior, lateral, and posterior vaginal-wall dilatation creates an anatomic basin for a theoretical seminal pool (see Part 2 of Chapter 6).

Full uterine elevation is not accomplished until the plateau phase of sexual response has been established. At this level of tension the nulliparous cervix may be elevated almost into the false pelvis. With the reaction of uterine elevation fully established, there is no further total organ response to effective sexual stimulation during terminal plateau phase or the orgasmic experience.

With onset of the resolution phase, the elevated uterus begins its return to the unstimulated resting position in the true pelvis. This return of the uterus from the false pelvis drops the cervix into the anatomically contrived seminal basin in the transcervical depths

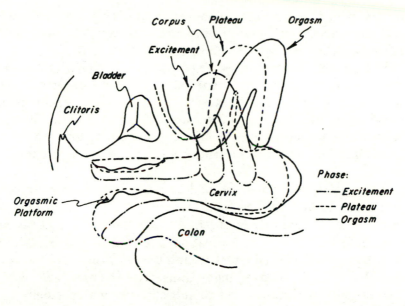

FIGURE 8-1

Uterine-elevation reaction. Composite of sexual response cycle.

of the vagina. Early in resolution uterine descent progresses rapidly. However, final return of the cervix to full apposition with the posterior vaginal wall frequently takes as long as five to ten minutes.

Currently there is no acceptable physiologic explanation for the phenomenon of uterine elevation in response to effective sexual stimulation. The uterus has been theorized to elevate in response to increased negative pressures in the abdominal cavity. In turn, the increase in negative pressure has been presumed to be the secondary result of the partial elevation and fixing of the diaphragms that usually develop in advanced excitement-phase levels of sexual tension. This theory does not explain the onset of uterine elevation early in the excitement phase, before any significant degree of hyperventilation develops. However, negative abdominal pressures may be a contributing factor to the final stages of uterine elevation late in the plateau phase.

The cardinal ligament and the plica rectouterina also have been suggested as possible contributors to the reaction of uterine elevation. Supporting this contention is the clinical observation that women with obstetrically traumatized uterine supports usually elevate the uterus much more slowly and with significantly reduced excursion, when compared to nulliparous women.

The generalized vasocongestive reaction in the pelvis (a constant response to sex tension increment) also has been suggested as contributing to uterine elevation through passive vasocongestion of the broad ligaments. As these major supports of the uterus fill with venous blood, there may be a resultant shortening and tensing of these structures which secondarily could contribute to uterine elevation.

Although there is no established physiologic reaction to account specifically for uterine elevation, there also is no question that the reaction occurs routinely in anteriorly placed uteri. The cervix has been observed through full elevation reactions hundreds of times with the use of artificial coital techniques, and the elevation reaction also has been observed many times directly during automanipulative episodes with the aid of an indwelling vaginal speculum.

The posteriorly positioned uterus (retroverted or retroflexed uterus) does not elevate from the true into the false pelvis in response to sexual stimuli. The posteriorly positioned uterus remains in the true pelvis, although vaginal-wall expansion does occur in the transcervical depth in the usual anterior, lateral, and posterior planes (see Part 1 of Chapter 6). There is no physiologic explanation for the lack of elevation of posteriorly placed uteri, particularly those that move freely on pelvic examination and obviously are not fixed in their posterior positions by adhesions or other forms of pelvic pathology.

The reactions of the individual components of the uterus (corpus and cervix) to advanced degrees of sexual stimulation constitute the second phase of total uterine response. The cervix and the corpus will be considered in sequence in discussing the reactive potential of the uterus to sexual stimulation.

CERVICAL RESPONSE

The production of vaginal lubrication in response to sexual stimulation has been identified with cervical secretory activity by many authors [8, 21, 30, 54, 67, 144, 219, 239, 295, 308, 318, 319]. During the past eleven years direct intravaginal observation of hundreds of complete cycles of female sexual response has been possible. With one exception, the cervix never has been observed to secrete material during any phase of the entire sexual response cycle. The exception was created by a study subject who lost an ovulatory-mucus plug from the external cervical os on the thirteenth day of what developed into a 28-day menstrual cycle. Although it is true that a thin, ovulatory-type mucoid material was discharged into the vaginal barrel, this secretory activity developed only after the responding woman was well into the plateau phase of the particular response cycle, long after full lubrication of the vaginal barrel had been established (see Part 1 of Chapter 6).

The only definitive response of the cervix to sexual stimulation develops with resolution-phase timing during the sexual cycle. A minimal dilatation of the external cervical os frequently has been observed. This specific cervical reaction, if it is to occur, develops immediately after an orgasmic experience. If the study subject does not achieve orgasmic tension release, the gaping of the external cervical os does not occur. When cervical os dilatation develops, 20 to 30 minutes of the resolution phase must intervene before a constrictive effect slowly closes the slightly patulous external os.

Positive identification of this reactive pattern has been confined to the nulligravid cervix. It well may occur in the parous individual, but the obstetric trauma resulting from passage of an infant's head and shoulders through the cervix obviates secure identification of this reaction subsequent to childbirth. The external cervical os normally is slightly patulous and the entrance to the cervical canal larger in the parous as opposed to the nulliparous woman. The development of dilatation of the external cervical os early in the resolution phase has been observed to parallel the intensity of the orgasmic response. Severe orgasmic experiences increase the frequency of identification of the patulous response of the external cervical os in the nulliparous woman.

It has been suggested that there may be clinical significance in the resolution-phase response of cervical gaping. This reaction theoretically increases the opportunity for spermatozoal access to the uterine cavity. While this physiologic possibility exists, the corpus contractile patterns to be described later in the chapter preclude an active role for cervical gaping in sperm migration. There is no question that gaping of the external cervical os, when it develops, may passively improve spermatozoal access to the cervical canal.

CORPUS RESPONSE

Clinical experiments have been conducted in attempts to evaluate uterine reactive potential and to find support for the concept that vasocongestion and myotonia are the physiologic bases of visceral reaction to sex tension increment.

MYOTONIA

Uterine physiology has been investigated with both intrauterine and abdominal electrode placements and acceptable physiologic recording techniques. Both investigative techniques have returned evidence of corpus irritability that increases from early in excitement to late in the plateau phase and resolves into an identifiable contraction pattern that has specific orgasmic-phase orientation and resolves with resolution-phase timing. These specific contraction patterns are repeated with each subjective orgasmic experience, varying in excursion (graphic representation of intensity) and duration of recordable result. This phase of the total program of basic physiologic investigation of human sexual response, involving not only target-organ but general body reactions, has not been established with sufficient statistical security for detailed presentation at this time. Suffice it to say that there is an identifiable recurrent pattern of uterine muscle contractility that is oriented specifically to the orgasmic phase of the female sexual cycle.

Specific uterine muscle contractile patterns do not develop unless the individual study subject undergoes an orgasmic experience that is recognizable both by trained observers and by the individual involved. Inevitably, the degree of excursion of recorded corpus contractive response parallels the study subject's subjective and the observers' objective evaluations of the physical and emotional intensity of the orgasmic experience. A uterine contraction pattern recorded with intrauterine electrodes during an orgasmic experience is presented in Figure 8-2. Typical corpus contractions of or-

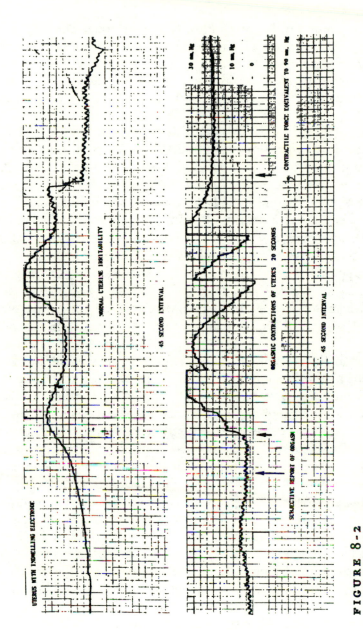

FIGURE 8-2

Uterine contraction in orgasm (intrauterine electrode).

117

gasmic response start in the fundus, progress through the midzone, and expire in the lower uterine segment. The contractile patterns are suggestive of those developed by the uterine musculature during the first stage of labor. However, the orgasmic-phase contractions are reduced in excursion, and contractile frequency is increased.

It will be noted (see Fig. 8-2) that fundal contractile patterns are initiated 2 to 4 seconds after subjective awareness of the onset of an orgasmic experience. Comparably, it has been observed that the orgasmic platform in the outer third of the vagina (see Part 1 of Chapter 6) may respond initially to the more severe levels of sexual tension (see "status orgasmus," Chapter 9) by contractile spasm, before this involuntary reaction is overcome by the regularly recurrent 0.8-second orgasmic contractions. There is little doubt that the female appreciates orgasmic response subjectively 2–4 seconds before she expresses this tension level physiologically. This lag between subjective appreciation and physiologic release may be comparable to the two stages of the male ejaculatory experience (see Chapter 14).

Although the number of experiments is not sufficient to allow an empirical position, it is current belief that the corpus contraction patterns initiated in response to automanipulative techniques are of greater intensity and duration than those resulting from coitally induced orgasmic experience. Certainly it is subjectively true that study subjects report that usually the experience with orgasm induced by masturbation is more intense than, although not necessarily as satisfying as, that resulting from coition (see Chapter 19).

Although initiated during orgasmic expression, corpus contractions continue in irregularly recurring patterns and with progressively reduced excursion as the individual woman is resolving from her orgasmic experience. There is at present no clinical explanation for the sensation of continued corpus contractions experienced at orgasm by study subjects with surgically ablated uteri. The probability of a pattern of conditioned response must be considered as a possible explanation of this subjective response.

The physiologic definition of uterine muscle contraction patterns during orgasm offers a possible explanation for the clinical complaints of cramping distress initiated during orgasmic response and experienced by many women (particularly multiparas). The orgasmic contractions of the corpus also have been reflected subjectively as painful stimuli in many instances by postmenopausal women (see Part 2 of Chapter 15).

VASOCONGESTION

A decade ago during the prostitute phase of the sex research program, investigators first were alerted to the extent that pelvic vasocongestion may be developed by long-continued sexual stimulation. One individual underwent repeated pelvic examinations during a six-and-a-half-hour working period and for six hours of observation thereafter. During the working period multiple coital exposures maintained the woman at excitement-phase levels of response. There were five subjective plateau-phase experiences superimposed on maintained excitement tension levels, but orgasmic relief was not experienced.

Toward the end of the working period the uterus was increased two to three times the unstimulated size; the broad ligaments thickened with venous congestion; the walls of the vaginal barrel were edematous and grossly engorged; and the major and minor labia were swollen two to three times normal size. Pelvic examinations and coital activity became increasingly painful toward the end of the six-and-a-half-hour working period.

During the six-hour observation period gross venous engorgement of the external and internal genitalia persisted—so much so, in fact, that the woman was irritable, emotionally disturbed, and could not sleep. She complained of pelvic fullness, pressure, cramping, moments of true pain, and a persistent, severe low backache. After the termination of the observation period automanipulation brought immediate relief from the subjective pelvic distress and the low backache. The objective findings also disappeared rapidly. Pelvic vasocongestion was reduced by an estimated 50 percent in five minutes and had disappeared completely ten minutes after the orgasmic experience.

Genital pain associated with long-maintained or continued sexual excitement has been described in the literature occasionally [2, 98, 191, 239]. However, the mechanism initiating the pelvic pain had

not been established prior to the clinical observations of the unresolved vasocongestion just described. Pain and pressure stimuli are
produced by severe unresolved vasocongestion of the pelvic target
organs. This clinical distress, although much more severe than that
classically described by Taylor as chronic passive congestion of the
pelvis [310–312], has the same pathologic orientation.

Results of the early prostitute observations instigated an experiment designed to investigate vasocongestive response in the female
pelvis during more moderate sexual activity. Over a four-year period
50 female study subjects cooperated with the program. They were
selected for ease of pelvic examination. Their ages ranged from 18
to 53 years. Thirty-one women were parous (13 had one child, 17
had two or three children, and one was a para 4); 19 women were
nulliparous. All women cooperated actively with the investigative
schedule on five different occasions. Two clinical observations were
conducted during the week immediately following a menstrual period, two clinical observations were conducted during the week immediately prior to anticipated menstruation, and one clinical experiment was conducted at the height of the menstrual flow (usually the second or third day of menstruation). Results returned
from the menstrual-flow observations will be discussed later in the
chapter.

The premenstrual and postmenstrual observation weeks included
both an automanipulative session and an active coital session. The
study subjects were stimulated to plateau-phase levels of sexual
tension. Automanipulation or coition was continued until orgasm
was judged imminent by the study subjects. Pelvic examinations
were conducted before onset of each stimulative session and immediately after arbitrary cessation of sexual stimulation late in the
plateau phase.

In every instance the 31 parous study subjects demonstrated a
significant increase in uterine size. Usually the uterus increased in
size from 50 to 100 percent over that described immediately prior
to the onset of sexual stimulation. In addition, those individuals
with any evidence of pelvic or labial varicosities demonstrated an
abnormal degree of venous engorgement of the broad ligaments.

Among the 19 nulliparous women the clinical results were not so
obvious. Only 7 nulliparous study subjects developed uterine size

increase of a clinical magnitude that was obvious during the pelvic examinations conducted late in the plateau phase of the sexual cycle. Four more nulliparous study subjects had suggestive uterine engorgement, but since there was a question of positive clinical identification they were listed as nonreactive. The remaining 8 women in the nulliparous group did not provide clinical evidence of uterine size increase at plateau-phase levels of sexual tension.

When the uterus obviously was enlarged, the deep vasocongestive response to sex tension increment developed both with automanipulative and coital stimulation. The longer sexually stimulative activity was continued before late plateau-phase levels of tension were achieved, the more severe was the deep vasocongestive response of the pelvic viscera.

Obviously, clinical examination of the pelvis is, at best, a crude determinant of pelvic vasocongestion. Yet, in women easy to examine pelvically a uterine size increase of from 50 to 100 percent developing from a known baseline within a twenty–thirty-minute interval should not be missed at routine examination. The clinical impression persists, after conducting these experiments, that in all pelves examined there probably was a significant increase in target-organ size due to vasocongestion. There was no question of this fact in the 31 parous study subjects; the 12 nulliparous subjects that were judged nonreactive may have been determined nonreactive only because the minimal degree of their pelvic visceral response could not be established by the crude technique of pelvic examination.

After the routine plateau-phase pelvic checks, the study subjects returned to sexual activity, and in approximately 75 percent of the occasions accomplished orgasmic-phase release of their sexual tensions. Examinations were conducted within a minute or two of the orgasmic experience, continued at five-minute intervals for a half-hour, and terminated at the hour interval. The nulliparous women lost all demonstrable vasocongestion from the pelvic viscera within 10 minutes after orgasm, while the multiparous study subjects needed from 10 to 20 minutes after orgasm before all clinical evidence of uterine vasocongestive size increase was dissipated. When orgasm was not accomplished, clinically obvious uterine size increase frequently remained for 30 to 60 minutes.

There no longer is any question that the uterus may and frequently does increase in size during a sexual response cycle. Particularly is this reaction clinically obvious when the excitement and plateau phases are extended in time sequence and the responding woman is parous. Thus, these experiments provide another example of deep vasocongestion in a target organ. The uterus under the influence of sex tension increment, together with the outer third of the vagina (orgasmic platform), the minor labia, and the breasts, reacts specifically to effective sexual stimulation by a marked vasocongestive increase in organ fluid content and consequently in organ size.

UTERINE RESPONSE AND SPERM MIGRATION

There have been numerous references to a sucking effect developed by the uterus and directed toward seminal-pool content [69, 103, 105, 126, 325]. Usually the concept is expressed that during orgasm the uterus develops some form of pressure and sucks the seminal fluid through the external cervical os into the cervical canal, and ultimately even into the endometrial cavity. Thus, in theory, uterine activity mechanically would shorten the transportation interval and the migratory distance for the spermatozoa elevated from the vaginal seminal pool by this reaction.

Evidence assembled during the past decade raises grave question as to the authenticity of this concept, so well established in biologic thinking. As described earlier in the chapter, corpus contractions start in the fundus, work down through the midzone, and terminate in the lower uterine segment. In other words, orgasmic uterine contractions are expulsive, not sucking or ingesting in character. Even if a negative pressure could be established in the uterus with its normal direct tubal connection into the abdominal cavity, corpus contractions would be expected to work from the midzone up toward the fundus in order to establish a sucking effect. In an attempt to evaluate this biologic concept, six study subjects were selected for active cooperation in a brief clinical investigation. They

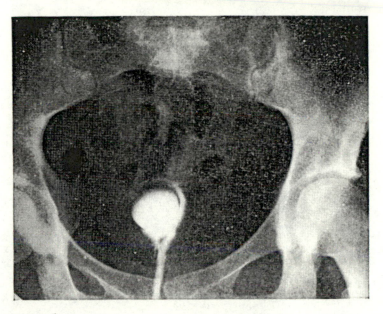

FIGURE 8-3

Check plate for cervical cap and radiopaque fluid. Note cervical immersion line.

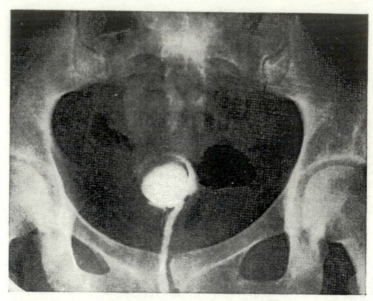

FIGURE 8-4

Cervix in contact with radiopaque material. Orgasmic phase.

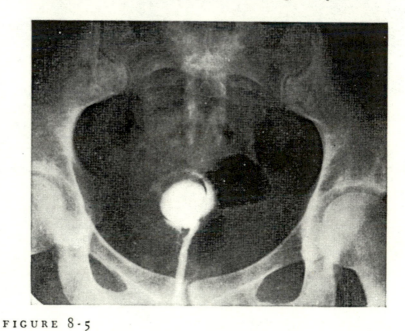

FIGURE 8-5

No evidence of sucking effect of uterus. Resolution phase (after 10 minutes).

were selected both for age and parity. There were two study subjects in their midtwenties, one nulliparous, the other multiparous. A similar selection was made of two study subjects in their midthirties, and finally, of two women in their early forties.

A reasonable facsimile of seminal-fluid content with relation to surface tension, specific gravity, and specific density was developed in a radiopaque substance with a liquid base. Since the time interval necessary for normal autolysis of seminal-fluid content is not present during active coition, every attempt was made to have the synthetic material correspond with immediately postejaculatory seminal fluid. Due to autolysis, repeated examination of seminal-fluid content provides constantly changing values. Baseline values could not be established for fresh seminal fluid with security and, therefore, no absolute parallel could be developed for the artificial media employed. Therefore, interpretive values within an admitted range of laboratory error arbitrarily were established for the substitute material employed. Average values of surface tension, specific density, and specific gravity were developed in the liquid-based radiopaque substance through the use of apple pectin.

The radiopaque substance was placed in a plastic cap and fitted over the cervix of each of the six participating study subjects. After the cap was placed, a radiographic check plate of the pelvis was taken to assure immersion of the cervix in the experimental material and to rule out the possibility of material spillage. Thereafter, radiographic plates were taken simultaneously with the orgasmic experience and after 10 minutes of the resolution phase. In none of the six individuals was there evidence of the slightest sucking effect on the media in the artificial seminal pool. Nor was there any evidence of the material in the cervical canal or the uterine endometrial cavity. All women were orgasmic during their experimental session.

The nulliparous woman in her middle twenties was selected as representative of the six study subjects. A radiogram taken immediately after the onset of automanipulation demonstrated the cervical cap in proper position and the line of cervical immersion (Fig. 8-3). This woman had an orgasmic experience within a five-minute time interval. A radiogram was taken during the orgasm (Fig. 8-4). The resolution-phase plate was taken 10 minutes after orgasmic response (Fig. 8-5).

In order to establish physiologically acceptable experimental conditions, all six individuals were evaluated at the time of expected ovulation. In each instance ovulation was established by the techniques of vaginal cytology [294]. While these clinical experiments are relatively uncontrolled, there still is no evidence to support the biologic concept of a sucking effect developed by the uterus during orgasmic experience.

From a purely physiologic point of view, the sucking concept cannot be supported by the mechanics of coition. If it is presumed that the uterus develops a sucking effect during orgasm in order to contribute actively to sperm migration, it must also be presumed that the external cervical os has immediate and relatively constant access to seminal-fluid content. During the woman's orgasmic experience the corpus and cervix are elevated far from the posterior vaginal floor. The penis frequently may intervene as a mechanical blocking agent, or the male may remain unejaculated at the moment of female orgasm. If ejaculation has occurred prior to female orgasm, the seminal-fluid content may be collecting in the anatomic basin on the posterior vaginal floor beneath the penis. Usually, the anatomy involved makes exposure of the cervical os to seminal-fluid content a virtual impossibility during active coition.

The biologic concept has been expressed that orgasm produces a negative pressure in the uterus which persists for significant lengths of time after active coition. If such a negative pressure or sucking effect were to develop, presumably a 10-minute resolution-phase radiogram would show some evidence of fluid migration into the cervical canal or the lower uterine segment. There was no such evidence in any of the six women checked. Thus it must be stated that there is no definitive evidence to date to support the concept of an active uterine role in aiding and abetting sperm migration from vaginal deposition sites.

UTERINE RESPONSE
DURING MENSTRUATION

Many cultures abound with interdiction against sexual activity during menstrual flow [41, 46, 70, 75, 79, 154, 266, 301]. While it certainly is not the province of this investigation to debate this point along cultural or religious lines, it does remain the province of the research to establish physiologic truth as opposed to cultural fiction. Frequently it has been presumed that coital activity during menstruation will lead to acute physical distress on the woman's part. During the past ten years no clinical evidence to support this concept has been established. In short, from a purely physiologic point of view, there is no contraindication to coition or automanipulation during menstruation. Obviously, if the female partner is subjectively distressed with the esthetics of coition during menstruation, is currently experiencing a heavy menstrual flow, or has obvious physical disability, she well may prefer to avoid coital connection. These subjective and objective concerns, when present, should be supported by full cooperation from her male partner.

There is the other side of the coin, however. Many women are interested in and desire sexual activity during their menstrual periods. Three hundred and thirty-one women were menstruating regularly during their active period of cooperation with the sex-research program. Each individual was questioned at length by interview teams with both sexes represented as to her concerns with or desires for sexual activity during her menstrual periods. Only 33 of these women objected to sexual activity during menstruation on a basis of religious or esthetic concerns. The remaining high percentage of study subjects vocalized no objection to or specific interest in sexual activity during their menstrual periods, providing they (1) were not at a peak of menstrual flow, (2) felt well, and (3) felt no esthetic aversion from the male partner. One hundred seventy-three of the 331 women expressed personal interest in sexual activity during menstruation, particularly during the last half of the flow period.

Also of interest was the fact that 43 women described the frequent use of automanipulative techniques with the onset of menstrual flow as a personal method contrived for relieving minor to major degrees of dysmenorrhea. These study subjects stated that severe orgasmic experience shortly after onset of menstruation increased the rate of flow, reduced pelvic cramping when present, and frequently relieved their menstrually associated backaches. Obviously, these women have learned a technique for the release of the cramping of excessive uterine irritability and the symptoms of pelvic vasocongestion when present at menstrual onset.

As stated previously, 50 women cooperated with investigation of uterine response to sexual stimulation during menstrual flow. Thirty-one of these women were parous and 19 were nulliparous individuals. At research request, the women subjectively selected the period of their heaviest flow for evaluation of uterine response. With a speculum placed in the vagina to provide full view of the cervix throughout the entire cycle, all 50 women achieved full orgasmic-phase tension release by automanipulative techniques. Seventeen of these women desired more than one orgasmic experience during the experimental sessions. During the terminal stages of orgasmic experience or within the first few seconds of the resolution phase, menstrual fluid could be observed spurting from the external cervical os under pressure. In many instances the pressure was so great that initial portions of the menstrual fluid actually were expelled from the vaginal barrel without contacting either blade of the speculum. It should be recalled that an indwelling speculum holds the cervix high and far from the vaginal outlet.

These observations provide further clinical evidence of uterine contractile response to effective sexual stimulation. A strong expelling force must be created in the corpus to extrude menstrual fluid under the degree of pressure evidenced in these 50 clinical observations. Since the menstrual flow was extruded in spurts rather than in continuous flow, the expulsive force can be presumed contractile rather than spastic in character. It should be recalled that corpus contractions initiated by orgasm have been recorded physiologically to start in the fundus and work toward the lower uterine segment. There now is objective clinical observation to support many women's subjective contention that sexual activity during menstrual flow markedly increases the flow on a temporary basis during an immediate postcoital or postmanipulative time sequence. Additionally, these observations support the concept of expulsive rather than ingestive reactions of the corpus to effective sexual stimulation.

9

THE FEMALE ORGASM

For the human female, orgasm is a psychophysiologic experience occurring within, and made meaningful by, a context of psychosocial influence. Physiologically, it is a brief episode of physical release from the vasocongestive and myotonic increment developed in response to sexual stimuli. Psychologically, it is subjective perception of a peak of physical reaction to sexual stimuli. The cycle of sexual response, with orgasm as the ultimate point in progression, generally is believed to develop from a drive of biologic-behavioral origin deeply integrated into the condition of human existence [55, 75, 142, 145, 196, 246, 279].

Where possible, material presented reflects consideration of three interacting areas of influence upon female orgasmic attainment previously recognized in attempts to understand and to interpret female sexual response: (1) physiologic (characteristic physical conditions and reactions during the peak of sex tension increment); (2) psychologic (psychosexual orientation and receptivity to orgasmic attainment); and (3) sociologic (cultural, environmental, and social factors influencing orgasmic incidence or ability) [12, 50, 65, 145, 290, 305, 322]. The quantitative and qualitative relationship of these factors appears totally variable between one woman's orgasmic experiences, and orgasm as it occurs in other women. Only baseline physiologic reactions and occasional individually characteristic modes of expression remain consistent from orgasm to orgasm, reflecting the human female's apparent tendency toward orientation of sexual expression to psychosocial demand.

Factual data pertaining to orgasm may be more meaningful when placed in clinical context. However, in order to provide a point of departure for nonsubjective interpretation of female orgasmic response, most of the material will be related to recognizable baselines of physiologic response and psychosocial patterns of sexual expression which can be duplicated within investigative context. General impression rather than statistical data will be reflected owing to the selected quality of the population and the research atmosphere to which the female study subjects have been exposed (see Chapter 2).

PHYSIOLOGIC FACTORS OF ORGASM

Female orgasmic experience can be visually identified as well as recorded by acceptable physiologic techniques. The primary requirement in objective identification of female orgasm is the knowledge that it is a total-body response with marked variation in reactive intensity and timing sequence. Previously, other observers have recognized and interpreted much of the reactive physiology of female orgasm [8, 18, 43, 44, 54, 69, 102, 116, 119, 169, 268, 278, 313, 314, 319]. However, definition and correlation of these reactions into an identifying pattern of orgasm per se has not been established.

At orgasm, the grimace and contortion of a woman's face graphically express the increment of myotonic tension throughout her entire body [144]. The muscles of the neck and the long muscles of the arms and legs usually contract into involuntary spasm. During coition in supine position the female's hands and feet voluntarily may be grasping her sexual partner. With absence of clutching interest or opportunity during coition or in solitary response to automanipulative techniques, the extremities may reflect involuntary carpopedal spasm. The striated muscles of the abdomen and the buttocks frequently are contracted voluntarily by women in conscious effort to elevate sexual tensions, particularly in an effort to break through from high plateau to orgasmic attainment (see Chapter 18).

The physiologic onset of orgasm is signaled by contractions of the target organs, starting with the orgasmic platform in the outer third of the vagina (see Part 1 of Chapter 6). This platform, created involuntarily by localized vasocongestion and myotonia, contracts with recordable rhythmicity as the tension increment is released.

The intercontractile intervals recur at o.8 second for the first three to six contractions, corresponding in timing sequence to the first few ejaculatory contractions (male orgasm) of the penis (see Part 1 of Chapter 12). The longer contractions continue, the more extended the intercontractile intervals. The number and intensity of orgasmic-platform contractions are direct measures of subjective severity and objective duration of the particular orgasmic experience. The correlation between platform contractions and subjective experience at orgasm has been corroborated by study subjects during thousands of cycles. Vaginal spasm and penile grasping reactions have been described many times in the clinical and nonprofessional literature [21, 147, 191, 228, 257, 271, 318]. Orgasmic-platform contractility provides an adequate physiologic explanation for these subjective concepts.

Contractions of the orgasmic platform provide visible manifestation of female orgasmic experience. To date, the precise mechanism whereby cortical, hormonal, or any unidentified influence may activate this and other orgasmic reactions has not been determined (perhaps by creating a trigger-point level of vasocongestive and myotonic increment).

Orgasmic contractions of the uterus have been recorded by both intrauterine and abdominally placed electrodes (see Chapter 8). Both techniques indicate that uterine contractions may have onset almost simultaneously with those of the orgasmic platform, but the contractive intensity of the uterine musculature is accumulated slowly and contractions are too irregular in recurrence and duration to allow pattern definition. Uterine contractions start in the fundus and work through the midzone to terminate in the lower uterine segment. With the exception of the factor of contractile excursion (indication of intensity), physiologic tracings of uterine orgasmic contractions resemble the patterns of first-stage labor contractions [107]. Uterine contractile intensity and duration vary widely from orgasm to orgasm. However, there is some early indication that both factors have a positive relation to the parity of the individual and the prior extent of her orgasmic experience, both incidental and cumulative.

Involuntary contractions of the external rectal sphincter also may develop during orgasm, although many women experience orgasm without evidencing sphincter contraction. When the contractions do occur, they parallel in timing sequence the initial intercontractile intervals of the orgasmic platform. The rectal-sphincter contractions usually terminate before those of the orgasmic platform.

The external urethral sphincter also may contract two or three times in an involuntary expression of orgasmic tension release. The contractions are without recordable rhythmicity and usually are confined to nulliparous premenopausal women.

The breasts evidence no specific response to the immediacy of orgasm. However, detumescence of the areolae immediately subsequent to orgasm is so rapid that its arbitrary assignment purely as a resolution-phase reaction has been cause for investigative concern. Often areolar detumescence is evident shortly after subjective report of orgasmic onset and usually develops simultaneously with the terminal contractions of the orgasmic platform. As a final stage of the rapid detumescent reaction, the areolae constrict into a corrugated state. The nipples remain erect and are turgid and quite rigid (the false-erection reaction).

Rapid detumescence of the vasocongested areolae, resulting in a constricted, corrugated appearance, occurs only with orgasm and is an obvious physical manifestation that provides for visual identification of female orgasmic experience. If orgasm does not occur areolar detumescence is a much slower process, corrugation does not develop, and the false-erection reaction of the nipples usually is reduced in intensity.

The sex flush, a maculopapular rash distributed superficially over the body surfaces, achieves its greatest intensity and its widest distribution at the moment of orgasmic expression. Subsequent to orgasmic experience, the sex flush disappears more rapidly than when resolving from plateau-phase levels of erotic tension.

From a cardiorespiratory point of view, orgasm is reflected by hyperventilation, with respiratory rates occasionally over 40 per minute. Tachycardia is a constant accompaniment of orgasmic experience, with cardiac rates running from 110 to beyond 180 beats per minute. Hypertension also is a constant finding. The systolic pressures are elevated by 30–80 mm. and diastolic pressures by 20–40 mm. Hg.

The clitoris, Bartholin's glands, and the major and minor labia are target organs for which no specific physiologic reactions to orgasmic-phase levels of sexual tension have been established.

Aside from ejaculation, there are two major areas of physiologic difference between female and male orgasmic expression. First, the female is capable of rapid return to orgasm immediately following an orgasmic experience, if restimulated before tensions have dropped below plateau-phase response levels. Second, the female is capable of maintaining an orgasmic experience for a relatively long period of time.

A rare reaction in the total of female orgasmic expression, but one that has been reduplicated in the laboratory on numerous occasions, has been termed *status orgasmus*. This physiologic state of stress is created either by a series of rapidly recurrent orgasmic experiences between which no recordable plateau-phase intervals can be demonstrated, or by a single, long-continued orgasmic episode. Subjective report, together with visual impression of involuntary variation in peripheral myotonia, suggests that the woman actually is ranging with extreme rapidity between successive orgasmic peaks and a baseline of advanced plateau-phase tension. Status orgasmus may last from 20 to more than 60 seconds, as reflected in Figure 9-1. The severe tachycardia (more than 180 per minute) and the long-maintained (43 seconds), rapidly recurring contractile patterns of the orgasmic platform are identified easily.

Of interest from both physiologic and psychologic points of view is the recorded evidence of an initial involuntary spasm of the orgasmic platform, developing before the regularly recurring contractions of orgasmic expression. As indicated in Figure 9-1, the study subject identified the onset of orgasm and vocalized this subjective experience before the onset of regularly recurrent contractions of the orgasmic platform. However, the initial spasm of the orgasmic platform developed parallel in timing sequence with the subjective identification of the orgasmic experience. To date, preliminary spasm of the orgasmic platform has been recorded only in situations of severe tension increment.

It is investigative impression that the inability to record initial spasm of the orgasmic platform in all orgasmic experiences well may reflect lack of effective experimental technique rather than unimpeachable physiologic fact. Subjectively, the identification of

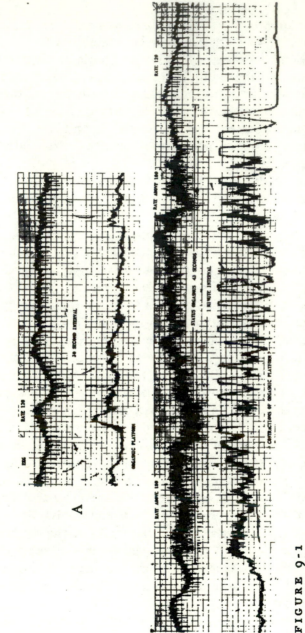

FIGURE 9-1

Status orgasmus (43 seconds): combination electrocardiogram and orgasmic platform recordings. (A) Late excitement phase: (*Top*) Heart rate at 120/min. (*Bottom*) Orgasmic platform: onset of irritability. (B) Orgasmic phase: (*Top*) Heart rate above 180/min. at peak. (*Bottom*) 25 regularly recurrent platform contractions.

initial spasm of the orgasmic platform is a constant factor in any full orgasmic experience. The subjective and objective correlation of orgasmic experience will be discussed later in the chapter.

No preliminary spastic contraction of the uterine musculature comparable to the initial spasm of the orgasmic platform has been recorded to date. However, the work is in its infancy, and such a preliminary spasm before onset of the regular, expulsive, fundal contractions may, in fact, exist and be recorded in the future.

The subjective identification of orgasmic expression by the human female simultaneously with the initial spasm of the orgasmic platform, but 2 to 4 seconds prior to onset of its regularly recurrent contractions, draws an interesting parallel with the human male's ejaculatory experience. When the secondary organs of reproduction contract (see Chapter 14), the male feels the ejaculation coming and can no longer control it, but there still is a 2- to 4-second interval before the seminal fluid appears at the urethral meatus under the pressure developed by penile expulsive contractions. Thus the male's psychosensory expression of ejaculatory inevitability may have counterpart in the female's subjective identification of orgasmic onset. The initial spasm of the orgasmic platform, before the platform and the uterus contract with regularity, may parallel the contractions of the prostate and, questionably, contractions of the seminal vesicles before onset of the regularly recurrent expulsive contractions of the penis.

Understandably, the maximum physiologic intensity of orgasmic response subjectively reported or objectively recorded has been achieved by self-regulated mechanical or automanipulative techniques. The next highest level of erotic intensity has resulted from partner manipulation, again with established or self-regulated methods, and the lowest intensity of target-organ response was achieved during coition.

While variations in the orgasmic intensity and duration of target-organ response have been recorded and related to modes of stimulation, there have been no recorded alterations in the basic orgasmic physiology. This finding lends support, at least in part, to many earlier concepts of orgasmic response [7, 16, 28, 29, 36, 51, 57, 64, 66, 76, 140, 144, 193, 194, 245, 256, 292, 315]. The fundamental physiology of orgasmic response remains the same whether the mode of stimulation is heterosexual or artificial coition or mechanical or automanipulative stimulation of the clitoral area, the breast, or any other selected erogenous zone. It follows that orgasm resulting from fantasy also would produce the same basic physiologic response patterns, although a woman capable of fantasying to orgasm has not been' available for inclusion in the research population. The ability of women to fantasy to orgasm has been reported by other investigators [21, 58, 144, 147, 152, 244, 267, 300].

With the specific anatomy of orgasmic-phase physiology reasonably established, the age-old practice of the human female of dissimulating has been made pointless. The obvious, rapid detumescence and corrugation of the areolae of the breasts and the definable contractions of the orgasmic platform in the outer third of the vagina remove any doubt as to whether the woman is pretending or experiencing orgasm. The severe vasocongestive reactions reflecting higher levels of sexual tension cannot be developed other than during involuntary response to sexual stimulation. For example, the transitory but obvious increase in nulliparous breast size, the sex flush, and the minor-labial sex skin reactions are all plateau-phase phenomena that develop only in response to effective sexual stimulation.

PSYCHOLOGIC   FACTORS   OF   ORGASM

It is well to restate from time to time the necessity for maintaining a concept of total involvement when any facet of human sexuality is to be considered. This is equally true when the study is directed to areas of psychologic influence upon orgasmic achievement [144, 232, 291].

Female orgasm, whether it is attained within the context of an interpersonal relationship (either heterosexual or homosexual) or by means of any combination of erotically stimulative activity and/ or fantasy, remains a potpourri of psychophysiologic conditions and social influence. Many theoretical as well as individually graphic accounts of the female experience at orgasm have been offered in the professional literature of many disciplines and are even more widespread in general publications. This vast amount of published quasiauthority depicts both objective and subjective female reaction to orgasm with almost every possible degree of accuracy and inaccuracy.

Without referring to the prior literature, a description of subjective response to orgasmic incidence has been compiled from reports of 487 women, given in the laboratory in the immediacy of the postorgasmic period, obtained through interview only, or developed from a combination of both sources. This composite is offered as a baseline for a concept of the psychologic aspects of the human female's orgasmic experience.

The consensus drawn from the multiple descriptions has established three distinct stages of woman's subjective progression through orgasm.

STAGE I

Orgasm has its onset with a sensation of suspension or stoppage. Lasting only an instant, the sensation is accompanied or followed immediately by an isolated thrust of intense sensual awareness, clitorally oriented, but radiating upward into the pelvis. Intensity ranging in degree from mild to shock level has been reported by many women within the context of their personal experience. A simultaneous loss of overall sensory acuity has been described as paralleling in degree the intensity and duration of the particular orgasmic episode. Loss of sensory acuity has been reviewed frequently in the literature [8, 30, 40, 69, 113, 114, 136, 144, 239, 251, 257, 271, 308, 319].

During the first stage of subjective progression in orgasm, the sensation of intense clitoral-pelvic awareness has been described by a number of women as occurring concomitantly with a sense of bearing down or expelling. Often a feeling of receptive opening was expressed. This last sensation was reported only by parous study subjects, a small number of whom expressed some concept of having an actual fluid emission or of expending in some concrete fashion. Previous male interpretation of these subjective reports may have resulted in the erroneous but widespread concept that female ejaculation is an integral part of female orgasmic expression [67, 93, 319].

Twelve women, all of whom have delivered babies on at least one occasion without anesthesia or analgesia, reported that during the second stage of labor they experienced a grossly intensified version of the sensations identified with this first stage of subjective progression through orgasm. Reports of this concept also have appeared from time to time in the literature [66, 241].

STAGE II

As the second stage of subjective progression through orgasm, a sensation of "suffusion of warmth," specifically pervading the pelvic area first and then spreading progressively throughout the body, was described by almost every woman with orgasmic experience.

STAGE III

Finally, as the third stage of subjective progression, a feeling of involuntary contraction with a specific focus in the vagina or lower pelvis was mentioned consistently. Frequently, the sensation was described as that of "pelvic throbbing."

Women with the facility to express sensate awareness frequently separated this final stage of subjective progression into two phases. The initial phase was expressed as contractile, followed immediately by a throbbing phase, with both sensations experienced as separate entities. The initial contractile feeling was described as localized vaginally, subsequently merging with the throbbing sensation which, though initially concentrated in the pelvis, was felt throughout the body. The "pelvic throbbing" sensation often was depicted as continuing until it became one with a sense of the pulse or heartbeat.

Only the two phases of this third stage of subjective progression during orgasm afforded positive correlation between subjective response and objective reaction. This correlation has been developed from a composite return of direct interrogation of female study subjects during investigative sessions. The phase of contractile sensation has been identified as paralleling in time sequence the recorded initial spasm of the orgasmic platform.

Regularly recurring orgasmic-platform contractions were appreciated subjectively as pulsating or throbbing sensations of the vagina. Although second-phase sensations of pulsation coincided with observable vaginal-platform contractions, consciousness of a pulsating sensation frequently continued beyond observable platform contractions. Finally this pelvic-throbbing sensation became one with a subjective awareness of tachycardia described frequently as feeling the heartbeat vaginally. Subjective awareness of orgasmic duration was somewhat dependent upon the degree of intensity of the specific orgasm.

Rectal-sphincter contraction also was described by some anatomically oriented or hypersensitive women as a specific entity during intense orgasmic response.

Observation supported by subjective report indicates that a relative norm of orgasmic intensity and duration is reflected by approximately five to eight vigorous contractions of the orgasmic platform. A level of eight to twelve contractions would be considered by observer and subject to be an intense physiologic experience. An orgasmic expression reflected by three to five contractions usually is reported by the responding female as being a "mild experience" unless the woman is postmenopausal (see Chapter 15). These physiologically recordable levels of orgasmic intensity never must be presumed arbitrarily to be a full or consistent measure of the subjective pleasure derived from individual orgasmic attainment [182, 197, 259].

Pregnancy (particularly during the second and, at times, the third trimester) has been noted to increase general sensitivity to the overall sensate effects of orgasm (see Part 2 of Chapter 10). To date, an increase in contractile intensity of the pregnant woman's orgasmic platform as compared to that in her nonpregnant state has not been corroborated by physiologic tracings. Orgasmic contractions of the uterus recorded during the second and third trimesters consistently have been reported as subjectively more intense sensations than those of nonpregnant response patterns. Of interest from an objective point of view is the fact that tonic spasm of the uterus develops in response to orgasmic stimulation and has been recorded during the third trimester of pregnancy.

SOCIOLOGIC   FACTORS   IN   ORGASM

In our culture, the human female's orgasmic attainment never has achieved the undeniable status afforded the male's ejaculation. While male orgasm (ejaculation) has the reproductive role in support of its perpetual acceptance, a comparable regard for female orgasm is still in limbo. Why has female orgasmic expression not been considered to be a reinforcement of woman's role as sexual partner and reproductive necessity? Neither totem, taboo, nor religious assignment seems to account completely for the force with which female orgasmic experience often is negated as a naturally occurring psychophysiologic response.

With orgasmic physiology established, the human female now has an undeniable opportunity to develop realistically her own sexual response levels. Disseminating this information enables the male partner to contribute to this development in support of an effective sexual relationship within the marital unit [62, 235]. The female's age-old foible of orgasmic pretense has been predicated upon the established concept that obvious female response increases the male's subjective pleasure during coital opportunity. With need for pretense removed, a sexually responding woman can stimulate effectively the interaction upon which both the man's and woman's psychosocial requirements are culturally so dependent for orgasmic facility.

Impression formed from eleven years of controlled observation suggests that psychosocially oriented patterns of sexual expression evolve specifically in response to developing social and life cycle demands. When continuity of study-subject cooperation permitted long-range observation and interrogation, it was noted that major changes in social baseline were accompanied by actual changes in sexual expression. For the female study subjects, changes involving social or life-cycle demands frequently resulted in a reorientation of sexual focus. This was manifest in alterations in desired areas of stimulation, preferred actions of partner, and reported fantasy [118, 144, 198]. Often variations in coital and masturbatory techniques were observed.

These alterations usually appeared gradually, although, depending upon the impact of the social change involved, there were occasions of sudden onset. To date, physiologically measurable intensity in orgasmic response has shown no specific parallel to onset or presence of these psychosocial influences. This may indicate that physiologic capacity, as influenced by purely biologic variations, remains a dominant factor in orgasmic intensity and facility [64, 76, 231, 316]. Reported levels of subjective pleasure in orgasm did, of course, parallel reports of desirable or undesirable social change.

It became evident that laboratory environment was not the determining factor in the success or failure of female study subjects' orgasmic attainment. Rather it was from previously established levels of sexual response that the individual female was able to cope with and adapt to a laboratory situation [76, 161, 170, 178, 221].

There were no particular personality trends toward high- or low-dominance individuals among the participating female research group. The women's personalities varied from the very shy through the agreeably independent, and histories reflected sexual-partner experience ranging from single to many. The ability to achieve orgasm in response to effective sexual stimulation was the only constant factor demonstrated by all active female participants. This observation might be considered to support the concept that sexual response to orgasm is the physiologic prerogative of most women, but its achievement in our culture may be more dependent upon psychosocial acceptance of sexuality than overtly aggressive behavior [17, 50, 57, 69, 100, 142, 194, 199, 300].

Many existing psychologic theories find support in the physiologic data emerging from this study [33, 81, 111, 166, 182, 233, 247]. However, it must be recalled that these data have been presented primarily as impression, due to the selectivity of the research group and, in many instances, the absence of a statistically significant number of recorded reactions. There always is great temptation to connect theory to considered fact, when subjective reports of the research population are placed as an overlay on the observed and recorded physiologic reactions. If recall by interrogated subjects of early sexual feeling and of manipulative activity, often to a remembered peak of experience, is to be given credence, sexual response well may be viewed as an instinctual activity arising from an undifferentiated sexual state. Although molded and transmitted genetically, sexual response, in this concept, would be subject to both immediate and continued learning processes [20, 71, 74, 82, 101, 160, 167, 181, 229, 332].

Unreported observations [131] suggest that infant sexual response as an undifferentiated state is not beyond possibility [82, 172]. Certainly, elaboration of sexual behavior in early childhood of less restricted cultures has been reported [32, 67, 76, 109, 155, 192, 228]. The development of sexual responsiveness to orgasmic level, identifiable subjectively, must be a cumulative result of interaction between the individual female's hereditary endowment and the psychosocial influence to which it is exposed. The element of time must be assumed to be a finally determining factor, as it accrues the experience of social and psychosexual maturation [112, 160, 180, 275].

A detailed psychosocial study of the research population cannot be presented within the framework of this text. Yet neither this book nor this chapter can be considered complete without emphasizing an acute awareness of the vital, certainly the primary influence, exerted by psychosocial factors upon human sexuality, particularly that of orgasmic attainment of the female [1, 53, 55, 76, 114, 145, 195, 313]. Although the basic physiology of female orgasm never would have evolved from behavioral theory or sociologic concept, it equally is obvious that physiologic detail is of value only when considered in relation to these theories and concepts. When completed, psychosocial evaluation of the study-subject population will be published in another book.

I O

PREGNANCY AND
SEXUAL RESPONSE

1. ANATOMY AND PHYSIOLOGY

Female sexuality as it may be enhanced or repressed by the state of pregnancy has been a subject of conjecture for generations. The taboos and sanctions dealing with this problem have been approached in number only by those which relate sexuality to menstrual flow. Many of these taboos and sanctions have been presumed based upon established biologic fact, when in reality they have developed from sources ranging from obscure speculation to that extreme of prejudice, restricted individual experience [14, 125, 171, 266, 280, 301, 309].

One hundred and eleven women aged 21 to 43 years agreed to cooperate with a subjective investigation of female sexual response as affected by pregnancy. Seventy-nine husbands of these women also cooperated with the program. The subjective material returned from team interviews will be presented in Part 2 of this chapter in a discussion format.

Greater value has been placed upon the objective results returned from a small group of six study subjects who cooperated with an anatomic and physiologic evaluation of sexual response during pregnancy and the postpartum period. The pertinent data relating to these study subjects are listed in Table 10-1. All women in this experimental group were married, and Subjects B, C, E, and F had taken active parts in various phases of the program before pregnancy. Study subjects B and E had cooperated with physiologic investigations of uterine contractile response to sexual stimuli prior to attempting pregnancy (see Chapter 8). The contractile patterns recorded from their nonpregnant uteri have been available for com-

141

TABLE 10-1
Pregnant Study Subjects

Subject Identification	Age at Conception	Gravidity and Parity *	Formal Education †	Study Subject Prior to Pregnancy
A	21	Grav. I Para 0	High school	No
B	21	Grav. III Para 1	High school	Yes
C	24	Grav. I Para 0	College	Yes
D	27	Grav. III Para 2	College	No
E	31	Grav. II Para 0	Postgraduate school	Yes
F	36	Grav. IV Para 3	High school	Yes

* Status with current pregnancy.
† Listing relates to reported matriculation (highest level).

parison with those returned from experiments conducted in uterine contractile physiology during the three trimesters of pregnancy.

Study subjects A and D first became active participants early in their pregnancies, A at seven weeks and D after eight and one-half weeks of amenorrhea.

Techniques of sexual stimulation have been natural and artificial coition and manual and mechanical automanipulation.

In describing the effects of a state of pregnancy upon female response to sexual tensions, no attempt will be made to present finite detail other than of target organs. The study subject group of six women is entirely too small to allow authoritative consideration of general body reactions. The target organs have been defined arbitrarily as the breasts and internal and external genitalia. Changes wrought in these organs by sexual stimulation superimposed upon alterations resultant from a state of pregnancy will be considered within the established framework of the four phases of the cycle of sexual response. Since stages of pregnancy also may influence target-organ response, physiologic alterations subsequent to sexual stimulation will be related to the three trimesters of pregnancy and the postpartum period.

THE BREASTS

In the human female's body some of the earliest changes reflecting pregnancy occur in the breasts. Tumescence develops in the areolae as an early indication of conceptive influence. The breasts rapidly increase in size during the first trimester due to significant increases in the vascular and glandular beds. Venous drainage patterns are defined clearly on breast surfaces early in the first trimester and continue through pregnancy and into the postpartum period.

The onset of tenderness along lateral breast surfaces and rapid size increase are the particular trademarks of the newly pregnant woman who has borne no children. When the nulliparous woman responds to sexual stimuli in the first trimester of her pregnancy, venous congestion of the breasts is more obvious than in a nonpregnant state. Hundreds of observations of nonpregnant women have established the fact that nulliparous breasts undergo a transitory 20–25 percent increase in size in response to plateau-phase levels of sexual tension (see Chapter 3). The same relative size increase usually develops in the nulliparous breast as a normal physiologic change by the end of the first trimester of pregnancy. When the vasocongestive reaction to sexual stimuli is superimposed upon the definitive increase in breast size resulting from a state of pregnancy, the nullipara's subjective complaint of severe breast tenderness during advanced stages of sexual tension early in pregnancy is understandable. Breast pain occasioned by sex tension increment frequently is localized in turgid nipples and engorged areolar elements.

During the second and third trimesters of pregnancy there usually is marked reduction in the nullipara's complaint of breast tenderness, whether it is solely the result of pregnancy or subsequent to the transitory venous congestion of superimposed sexual tension. By the time the conical nursing-shape characteristic of late third trimester has been achieved, breast volume has been increased by approximately one-third compared to that in the nonpregnant state. High levels of sexual tension frequently do not provide further obvious increase in breast size at this stage of pregnancy. However, reactions of nipple erection and areolar tumescence remain constant through all three trimesters of pregnancy.

During the second or third month of the postpartum period, breast reaction to sexual stimuli is related to the nursing phenomenon. If the normal postpartum milk production has been depressed artificially through hormonal control or by constrictive breast binding, there is little physiologic response of the breasts other than nipple erection, even to plateau-phase levels of sexual tension. The woman with suppressed milk production may be six months past her delivery date before any definitive vasocongestive reaction can be observed in the breasts in response to sexual stimulation.

Nursing breasts, normally increased in volume, do not demonstrate a consistent size increase even at plateau-phase levels of sexual tension. However, the nursing mother responding to sex tension increment frequently initiates an unusual reaction pattern. Many women lose breast milk in uncontrolled spurts when responding to sexual stimuli. Milk has been observed to run from both nipples simultaneously during and immediately subsequent to an orgasmic experience. Usually full flow of milk is restricted from the breast more recently nursed. Involuntary loss of milk control has been observed during both coital and automanipulative activity.

Since only six women cooperated actively during and immediately after their pregnancies, no suggestion of statistical significance can be derived from clinical observation. Three of the six women did nurse after their deliveries, two of them for a four-month and one for a six-month period. Two of these women frequently demonstrated involuntary loss of milk control during orgasmic experience. They also reported the experience of similar reaction patterns outside the laboratory. The third study subject in the nursing group did not lose milk control during orgasmic response either during episodes of active cooperation with the research program or in private reactive experience.

THE GENITAL ORGANS

A state of pregnancy markedly increases the vascularity of the pelvic viscera. Obviously, the mechanism of fetal support creates gross vasoconcentration in the female pelvis. Any superimposed physiologic response to sexual stimulation even further increases this massive pelvic vasocongestion.

All six of the study subjects became consciously aware of increased levels of sexual tension toward the end of the first or during the early stages of the second trimester. The sex tension increment reached extremely high levels during the second and continued well into the third trimesters of their pregnancies. Four of the six women described occasional cramping and aching in the midline of the lower abdomen during and immediately subsequent to orgasmic experience in the first trimester of pregnancy. Two women also complained of low backache as a residual of the orgasmic cramping episodes. Although none of the six women spotted vaginally or developed any clinical threat of pregnancy wastage, the subjective awareness of increased uterine irritability subsequent to effective sexual stimulation was of particular moment.

During the second trimester, all six study subjects described strong sexual drives marked by increased interest both in coital and manipulative activity and were observed in fulminating orgasmic experiences. Two subjects who had never been multiorgasmic in prior sexual experience described and demonstrated the onset of this high-tension response for the first time during the second trimester of their pregnancies. The other four study subjects have positive histories for multiorgasmic response both in pregnant and nonpregnant states. Discussions of reactions of external and internal genitalia to sex tension increment during states of pregnancy will be restricted primarily to variations from the norms of physiologic response established for the nonpregnant state (see Part 1 of Chapter 6).

EXCITEMENT PHASE

The human female's external genitalia and the internal organs of reproduction alter significantly during pregnancy, when compared to their norms in the nonpregnant state. Variations in pelvic viscera, in reaction to pregnancy, primarily are related to intensity of generalized pelvic vasocongestion. Therefore, it is understandable that the vasocongestive response of the internal and external genitalia to sexual stimuli is a much more significant factor in pregnancy than is the development of myotonia.

During the excitement phase the reaction of the labia majora in the nullipara follows the usual patterns (described in Chapter 4). For multiparous women, however, there is a tendency for the major labia to be excessively engorged with blood and frequently quite edematous. Although they undergo involuntary lateral withdrawal from the vaginal outlet in the routine mounting invitation, the elevation and flattening reactions of the major labia usually are absent after the first trimester of pregnancy. This restriction in reactive potential probably results from a marked increase in vasocongestion of the entire pelvis associated with the pregnant state.

In response to excitement-phase levels of sexual tension, the minor labia usually become markedly engorged and at least two to three times enlarged. This vasocongestive response to sex tension increment continues throughout the first two trimesters of pregnancy. In the third trimester the minor labia normally are so chronically engorged with both venous blood and interstitial edema fluid that further specific distention subsequent to sex tension influence is difficult to demonstrate.

Toward the end of the first trimester all six study subjects described a definitive increase in the production of vaginal lubrication that continued throughout pregnancy. Since four of the six women had worked with the sex-research program prior to pregnancy, objective measurement of lubrication production was possible. Vaginal lubrication developed more rapidly and more extensively for these four women during pregnancy than had been their pattern in non-pregnant response to sexual stimulation. Usually, pregnancy-oriented increase in lubrication production was greater in the multiparous subjects as opposed to those women carrying their first term pregnancies.

All six study subjects complained of a light mucoid discharge that became apparent toward the end of the first trimester and continued throughout pregnancy. It may be that the gross venous engorgement of the vaginal barrel accompanying states of pregnancy provides such an increase in vaginal lubrication that there is more or less constant, low-grade production of this material, even when the individual is not exposed consciously to sexual stimuli. Such a reaction would be presumed the result of pelvic venous congestion contributing to the transudate-like product of vaginal lubrication.

By the end of the first trimester of pregnancy all uteri normally are so enlarged that they have become abdominal organs. This physiologic distention and elevation of the pregnant uterus includes those that are normally retroverted or retroflexed. One subject had a retroverted uterus that moved forward and elevated to become an abdominal organ by the end of the first trimester of pregnancy. After the uterus is elevated into the abdomen, vaginal expansion and distention in response to sexual stimuli continues in the same fashion as in the nonpregnant state. The exception is that the definitive tenting phenomenon in the transcervical vaginal depth, which occurs subsequent to excitement-phase uterine elevation in the non-pregnant state, cannot be demonstrated. Once the pregnant uterus becomes an abdominal organ, there is a normal physiologic "tenting" which develops in the anterior vaginal wall to such a degree that further direct response to sexual stimuli cannot be established with security.

There has been no evidence of secretory activity at the cervical os during sexual excitation in any of the six pregnant study subjects during the three trimesters of pregnancy.

PLATEAU PHASE

The minor labial sex-skin reaction occurs routinely in every female, pregnant or nonpregnant, if orgasm is to follow.

As plateau-phase tension levels are established, a marked venous engorgement of the outer third of the vagina normally develops as the orgasmic platform (see Part 1 of Chapter 6). In the nulliparous pregnant study subjects, localized vaginal engorgement became so severe when excitement or plateau phases were prolonged significantly that 75 percent of the vaginal lumen was obtunded by this massive vasocongestion. For the multiparous pregnant study subjects, the orgasmic platform developed to such an extent that frequently the vaginal barrel appeared completely obtunded, with the lateral vaginal walls meeting in the midline in severe vasocongestive response to sexual tensions. The more advanced the state of pregnancy, the more severe the venous engorgement of the entire vaginal barrel, and the more advanced the secondary development of the orgasmic platform in response to sexual stimulation.

The uterine-elevation reaction (see Chapter 8) cannot be demonstrated once the uterus becomes an abdominal organ. Uterine vasocongestive increase in size or vasocongestion of the broad ligaments cannot be identified as specific reactions to sex tension increment after pregnancy has been established. This does not mean that these reactions do not occur to varying degrees. The gross expansion of the uterus and the passive vasocongestion of the broad ligaments associated with normal pregnancy (particularly during the second and third trimesters) preclude identifying these reactions with the crude clinical techniques currently available.

ORGASM

Orgasmic-platform contractions can be identified as specific physiologic evidence of orgasmic experience during both the first and second trimesters of pregnancy. During the third trimester the orgasmic platform in the outer third of the vagina may be so congested with venous blood and the entire vaginal barrel so edematous that intensity of contractions during orgasmic experience appear minimal to direct observation. Although the woman subjectively feels contractile response of the orgasmic platform, the entire area may be so overdistended that objective evidence of contractile efficiency is reduced markedly.

During the third trimester of pregnancy, particularly during the last few weeks before term, the uterus instead of contracting regularly during orgasmic experience may go into tonic spasm. Spastic uterine contractions have been observed occasionally in the laboratory and timed to continue for as long as one minute in response to orgasmic stimulation. Listening to the fetal heart tones at this time may return evidence of a slowed heart rate, but this reaction is transitory in character. No further evidence of fetal distress has been demonstrated. Two study subjects evidenced regularly recurring uterine contractions for as long as a half-hour after orgasmic experience in the last month of pregnancy. Obviously, the uterus is normally highly irritable at this time, particularly in the nulliparous woman. When regularly recurrent uterine contractions associated with orgasmic experience are superimposed upon the basic uterine irritability of the last stages of pregnancy, a spastic uterine contraction may be the result.

RESOLUTION PHASE

Resolution-phase reaction during pregnancy differs severely from that in the nonpregnant state, in that the vasocongested pelvis frequently is not relieved completely with orgasmic experience. The further pregnancy progresses, the less effective is vasocongestive disbursement subsequent to orgasm. Direct observation of the six pregnant study subjects revealed continued major and minor labial engorgement and a congested vaginal barrel that included residuals of the orgasmic platform, as opposed to the usual complete disbursement of this localized vasocongestion subsequent to orgasm in the nonpregnant state.

In resolution during the second trimester, it may take from 10 to 15 minutes after orgasm for the increased labial and vaginal-barrel vasocongestion developed by sex tension increment to be lost from the primigravid pelvis, and 30 or 45 minutes from the multiparous pelvis. As previously stated, during the third trimester vasoconcentration in the pelvis may not be relieved completely regardless of the severity of the orgasmic experience. This residual vasocongestion often is subjectively translated into a continuation of sexual stimulation.

The lack of clinical relief of chronic pelvic vasocongestion subsequent to orgasmic expression may account for the fact that during the second and third trimesters of pregnancy the six study subjects noticed subjectively higher sexual tension levels than they had encountered in nonpregnant states. The study subjects repeatedly stated that orgasmic experience, although objectively most severe and subjectively quite satisfying, did not relieve their sexual tension levels for any significant length of time. Obviously, their comparative focus was subjective recall of comparable orgasmic response situations in nonpregnant states.

The massive vasocongestion of the pelvic viscera associated with a state of pregnancy further is increased by sex tension increment even late into the third trimester of pregnancy. Postorgasmic disbursement of the pelvic vasoconcentration, a normal resolution-phase reaction in nonpregnant states, is slowed and usually transitory during pregnancy. Residual pelvic vasocongestion, together with the pelvic pressures resultant from second- and third-trimester uteri, may account for the high levels of maintained sexual tensions frequently described for these stages of pregnancy.

POSTPARTUM PHYSIOLOGY

All six study subjects had uncomplicated deliveries, all babies were reported to be in good physical condition, and the six subjects rejoined the investigative program between the fourth and fifth weeks after delivery. As noted earlier, three subjects nursed through the fourth postpartum month. All subjects were reevaluated three times after delivery: when they rejoined the program, between the sixth and eighth postpartum weeks, and at the end of the third postpartum month. A pelvic check with the first evaluation at four to five weeks showed the episiotomies to be well healed, the cervices closed, and the uteri still abdominal organs. The nursing subjects, as would be expected, had smaller, better involuted uteri than the nonnursing women. From the first evaluation at four to five weeks to the last observations at the third postpartum month, major physiologic changes developed in the pelves.

Although four of the six study subjects reported significant levels of eroticism at the first check, the physiologic reactions of their target organs were reduced both in rapidity and intensity of response. Vasocongestive reactions of the major and minor labia were mature, once they developed, but they frequently were delayed in development well into the plateau phase. Vaginal lubrication developed slowly and in reduced quantity. Vaginal distention in the inner two-thirds of the barrel also was reduced in rapidity of development and in the degree of excursion when compared to previously established patterns of reactivity. Under direct observation the walls of the vagina were quite thin at this time. Normal rugal patterns were flattened or absent and the vagina was light pink in color and appeared almost senile to direct observation. Particularly was this steroid-starvation pattern true for the three nursing mothers. Since the uteri were still abdominal organs and a residual of pregnancy-incurred venous congestion remained in the broad ligaments, no true clinical picture of supravaginal vasocongestive response to sex tension increment could be obtained during these examinations.

At plateau-phase tension levels the orgasmic platform developed in the outer third of the vagina, but there was significant reduction in the extent of reaction. During the last trimester of pregnancy, 75 to 100 percent of the vaginal lumen had been obtunded by the development of the orgasmic platform; at the four-to-five-week postdelivery check not more than one-third of the vaginal barrel was obtunded by the orgasmic platform immediately prior to orgasmic experience. The sex-skin reaction of the minor labia was present in all women immediately prior to orgasm, but there was significant reduction in the vividness of the color change.

With orgasm, contractions of the orgasmic platform were reduced in intensity and duration of recurrence. Although the study subjects reported subjective satisfaction from orgasmic experience, the orgasmic-platform contractions were reduced markedly in physiologic intensity during the actual orgasmic experiences.

At the six-to-eight-week check there was little variation from the findings during the first examinations.

Early postpartum response to sex tension increment has provided an exception to the general rule that physiologic response patterns parallel in intensity the psychosexual tension levels. Through the first six to eight postpartum weeks, sexual tensions frequently were described at nonpregnant levels, particularly among the nursing mothers, but intensity and duration of physiologic response during coital and manipulative opportunity were diminished. Thus there is a suggestion that states of steroid starvation may have more effect on physiologic patterns of performance than on psychologic levels of tension.

At the end of the third postpartum month an entirely different picture was presented. All six study subjects gave evidence of return of ovarian hormone production although the nonnursing group was ahead of the three nursing mothers. Vaginal rugal patterns were reestablished, and uteri had returned to normal pelvic positioning. The major and minor labia responded readily to sexual stimuli following response patterns established for women in a nonpregnant state (see Part 1 of Chapter 6). Lubrication developed in expected quantity and with usual rapidity. The vaginal barrels expanded at the transcervical depth and extended in length in the nonpregnant manner.

In response to plateau-phase tension levels, uterine elevation and vasoconcentration of the broad ligaments could be identified. Uterine size increase could not be determined with security. The orgasmic platform developed to normal degree, obtunding approximately 50 percent of the vaginal outlet. The sex-skin reactions of the minor labia again developed a vivid coloration in the immediate preorgasmic period.

With orgasm, the orgasmic platforms evidenced increased contractile intensity and the contractions recurred from eight to twelve times, well within usual patterns of nonpregnant response.

Subjectively, the study subjects could not define significant difference between the orgasmic experiences of the three-month check as opposed to those developed during the four-to-five-week check. Physiologically, however, there was no question of increased intensity and duration of the third month's experiences as opposed to those developed four to five weeks after delivery.

2 .   CLINICAL   CONSIDERATIONS

The pregnancy year (three trimesters and the immediate postpartum period) contributes to elevations and depressions of both male and female sexuality that represent excursions well beyond the response levels usually encountered in nonpregnant states.

Eroticism in pregnancy has not been investigated to a degree sufficient to establish response patterns acceptable as baselines by either the biologic or behavioral disciplines. In an attempt to highlight this wide gap in the knowledge of human sexual response, limited studies with both objective and subjective focus have been conducted. The physiology of human sexual response as affected by pregnancy has been presented in Part 1 of this chapter. The subjective concerns of the pregnant woman and the involved man in relating a state of pregnancy to sexual response are of current interest. This discussion provides a baseline from which a definitive study of human sexual behavior during pregnancy may be developed.

Material returned from this phase of the investigation must be accepted at the level of clinical impression rather than considered as statistically suggestive or presumed established fact. The number of women interrogated in depth is too restricted and the sample too biased to represent an adequate cross-section of the population. In presenting material, clinical import will be given precedent over subjective report. A more definitive study of pregnancy and sexual response will be presented in the near future.

Interrogative opportunity was solicited from 113 pregnant women. Of the women originally approached, 111, aged 21 to 43 years, agreed to cooperate with a subjective investigation of female eroticism as affected by pregnancy. During the first trimester initial interviews were scheduled toward the end of the second month. In the second trimester the sixth month was the review month, and for the third trimester review was conducted at the end of the eighth month of pregnancy. The postpartum review was held in the third month after delivery. The initial interview concentrated upon psychosocial, sexual, and medical backgrounds. The review periods primarily were sexually focused. All interviews were conducted by a team of both sexes.

Subjective material has fallen into patterns that relate to age, parity, trimester of pregnancy, postpartum period, current state of health, social pressures, and, of course, successful termination of pregnancy.

As shown in Table 10-2, the 111 pregnant women have been grouped by age and parity. A total of 43 of these women, 7 of whom were unmarried, were primigravidas and expected to carry to full term. Thirty-seven women were carrying a second pregnancy to term; 2 of these women were unmarried. Finally, 24 women

TABLE 10-2

*Ages and Parity of 111 Pregnant Women
Interviewed*

| | Ages (Yr.) | | | |
Parity	21–30	31–40	41–43	Totals
0	35	8	0	43
1	25	11	1	37
2	14	8	2	24
3	2	3	2	7
Totals	76	30	5	111

were attempting to carry a third pregnancy, and 7 a fourth, to full term. There were 76 women between 21 and 30 years old; 30 women between 31 and 40; and 5 women were between 40 and 43 years of age at the onset of the current pregnancy.

The statistics of previous pregnancy wastage and current pregnancy conclusion in the 111 women are listed in Table 10-3. Three of the 43 women attempting their first full-term pregnancy aborted toward the end of the first trimester, and 1 woman became severely toxemic and lost a stillborn infant three weeks from term. Three of the 37 attempting a second full-term pregnancy aborted before the end of the first trimester, and 1 miscarried at the fifth month. Finally, 2 of the 24 women attempting a third full-term pregnancy aborted before the end of the first trimester, 1 woman miscarried, and 1 lost an infant from congenital malformations and infection approximately one month postdelivery. None of the 7 women attempting a fourth pregnancy had obstetric difficulty.

In brief, of the total of 111 women interviewed during the second month in the first trimester, 8 women lost pregnancies shortly after their initial interview and 2 women lost pregnancies during their second trimester. This pregnancy wastage brought to 101 the number of women cooperating throughout the interrogation (included among the 101 women were the 2 women with pregnancy loss at term and in the postpartum period).

TABLE 10-3
Pregnancy Wastage in 111 Pregnant Women Interviewed

		No. Fetal Deaths							
		Previous Conceptions				Current Conception			
		1st Trimester		2nd Trimester Miscarriage	3rd Trimester: Delivery and Postpartum	1st Trimester		2nd Trimester Miscarriage	3rd Trimester: Delivery and Postpartum
Parity	No.	Abortion	C-AB*			Abortion	C-AB*		
0	43	12	7	2	0	3	0	0	1
1	37	7	3	1	2	3	0	1	0
2	24	6	2	0	2	2	0	1	1
3	7	2	0	0	0	0	0	0	0
Totals	111	27	12	3	4	8	0	2	2

* C-AB = Criminal abortion.

Seventy-nine husbands offered cooperation with team interrogation, and did so at the end of the third postpartum month after all review with their wives had been completed. Nine of the 101 women were unmarried. Thus 79 of a possible 92 husbands cooperated with the investigation. The age among the men interrogated ranged from 25 to 49 and averaged 31 years, 7 months. Formal education averaged three years and two months of college. The educational range was from incomplete high-school attendance to postgraduate degree.

The subjective material returned from interrogation of the cooperating women relates primarily to parity and trimester of pregnancy and will be presented in the framework of these arbitrary standards. The average age of the 101 women was 27 years, 8 months, and the average level of formal education was two years and three months of college exposure. As with the husbands, the actual range of formal education varied from incomplete high-school attendance to postgraduate degree.

Subjective material returned from repeated interviews with the 6 study subjects who cooperated with the physiologic investigation of sexual response in pregnancy (see Part 1 of this chapter) is not included in the material contributed by the 101 women who cooperated only at interview level.

FIRST TRIMESTER

During the first trimester, great variation was reported in levels of eroticism and effectiveness of sexual performance among the women interviewed. Reports ranged from voluntary rejection of all physical forms of sexual activity during the entire pregnancy to deliberate prostitution. In essence, the first-trimester response related closely to parity and to social stratum.

NULLIPAROUS WOMEN

Of the 43 women who were attempting a first, full-term pregnancy, 33 reported reduction in sexual tensions and in effectiveness of sexual performance. Many of these women were contending with nausea, and all were affected with sleepiness and symptoms of chronic fatigue.

Twenty-six of the 43 nulliparous women reported fear of injury to the conceptus (frequently not vocalized to their partner) as affecting the freedom of their physical response in coital activity during the first trimester. Two of these women had been urged to avoid coition during the first trimester by their physicians.

There were 7 unmarried women in the sample who were undergoing their first pregnancy. Two of these women continued coital connection throughout pregnancy and the postpartum period with the man who reportedly fathered the child. Three of the 7 women had occasional coital opportunities during the pregnancy. One girl rejected both intercourse and automanipulation for the duration of the pregnancy after a diagnosis was established, and one woman turned to prostitution during her first trimester, admittedly for financial gain, but with the expressed hope that excessive sexual activity early in the pregnancy would induce an abortion. Each of these 7 women reported little or no eroticism during their first trimester of pregnancy. They were overwhelmed with the social aspects of their circumstance. Concern for their financial security, plans for the baby, and concern for their own personal care during the pregnancy weighed heavily during this trimester. These socioeconomic concerns also were expressed by the two women with the semipermanent male-partner connections.

Among the remaining 10 members of the nulliparous group, 6 described no apparent change in their sexual interest or effectiveness of their sexual performance. Only 4 nulliparous women reported a significant increase in sexual interest and elevated demand for increased rate of performance immediately after the diagnosis of pregnancy was established.

PAROUS WOMEN

As a group, the 68 parous women generally noted very little change in their levels of sexual interest or effectiveness of performance during the first trimesters of their pregnancies as opposed to their recalled levels of response for a three-month period immediately prior to conception. Exceptions to this statement were

7 women who were involved with nausea and vomiting of pregnancy. All of these women described similar gastrointestinal involvement and similar marked loss of sexual interest during the first trimesters of previous pregnancies. Only 4 of 68 women reported an increase in sexual drive or improved sexual performance that was apparent by the end of the second month of pregnancy when the interrogations were conducted.

SECOND TRIMESTER

During the second trimester, sexual patterns generally reflected a marked increase in eroticism and effectiveness of performance regardless of the parity or ages of the women interrogated. This evidence of elevated sexuality was reported by the women not only as interest in sexual encounter but also as planning for sexual encounter, fantasy of sexual encounter, and sex-dream content. There also was an increased demand for a considered effectiveness of sexual performance.

Of the 101 women reviewed toward the end of the second trimester, 82 described a significant improvement in basic sexuality not only over that recalled from the first trimester of pregnancy but well beyond their concept of previously established norms of performance in nonpregnant states.

Of the 19 women who described no improvement in sexual interest or performance 11 were nulliparous and 8 were parous. Of these, 4 women (3 nulliparous) described extremely low levels of sexual interest in their past histories, had never been orgasmic, and could not define change in sexual interest or demand during their entire pregnancies. These women also had described themselves as sexually unaffected by pregnancy during interrogation in the first trimester.

Two women in the parous group were having their second illegitimate pregnancies. Both of these women were living in their concept of common-law marriage. They described marked increase in sexual interest and performance during the second trimester.

Five of the 7 women illegitimately pregnant with their first pregnancy had no opportunity for regular coition. However, 4 of these women described marked increase in sexuality. Although socially denied the opportunity of regular coition, they masturbated with a greatly increased frequency as compared not only to first-trimester activity but also to their pattern in the nonpregnant state. Their socioeconomic concerns were relatively under control, and their personal fears, if not dispelled, at least were neutralized. Among these 4 women was the girl who had prostituted herself during the first trimester of pregnancy. Although currently denied social opportunity for regularly recurring intercourse, she described marked subjective sexual drive for the first time and multiorgasmic performance on coital occasion. In addition she instituted auto-manipulative techniques on a regularly recurring basis for the first time in her life.

One girl, as previously described, rejected any form of sexual performance throughout her entire pregnancy and the postpartum period. She did describe constantly recurring erotic dreams and occasions of erotic fantasy during the second trimester.

THIRD TRIMESTER

Interviews were conducted approximately one month before the estimated date of confinement. In the nulliparous group there was a significant reduction in coital frequency compared to reported incidence in the second trimester. However, there was the major restraint of medical intervention. Intercourse was contradicted by medical advice for 31 of the 40 nulliparous women who attained the third trimester of pregnancy. Continence was prescribed for periods varying from four weeks to three months prior to delivery. There also were multiple reports of a variety of somatic complaints. The sleepiness of the first trimester returned, and the second-trimester symptoms of irritability, abdominal fullness, pelvic tension, and backache increased in severity. Although admittedly strongly influenced by medical restriction, 33 of the 40 nulliparous women reported that they personally gradually lost interest in sexual activity during the third trimester.

Intercourse also was contradicted by medical authority for 46 of the 61 parous women reviewed during the third trimester of their pregnancies. Again the period of coital restriction ranged from three months to four weeks before the estimated date of confinement. Forty-one of the 61 members of this parity group described a significant reduction in eroticism and frequency of sexual performance as the estimated date of confinement approached. The subjective loss of sexual tension was related by parous women to exhausting physical demands more than to medical influence. Frequently parous women admitted that contending with existing children and their own physical distresses magnified by the advanced stage of pregnancy kept them in such a state of chronic exhaustion that they seldom sought the opportunity for active sexual performance. When directly approached, however, their effectiveness in and capacity for sexual performance frequently surprised the women themselves.

Sixty-eight of the 77 women for whom coition was interdicted medically expressed concern with the prescribed period of sexual continence and its possible effect upon their husbands' sexual requirements. Seventeen of the nulliparous and 32 of the parous women reported that they made deliberate attempts to relieve their husbands during the period of prescribed continence.

Of interest were the observations of 8 nulliparas and 12 multiparas relating to husbands' loss of sexual interest in them personally. They stated that their husbands' withdrawal from sexual encounter had onset late in the second or early in the third trimester of pregnancy. By this time they had become obviously pregnant, with large abdomen, swollen face, thickened ankles, and all of the other gross physical signs of approaching confinement. They expressed concern over this apparent lack of male interest in their physical being, and the fear that the current rejection might have some degree of permanent residual. All but 3 women thought that their husbands avoided them on the basis of (1) their physical appearance, (2) concern for their personal comfort, or (3) fear of injuring the fetus, and they were content to wait for the release of delivery. The 3 women expressed specific knowledge of their husbands' interest in other sexual outlets at this time.

POSTPARTUM PERIOD

Postpartum interviews were conducted in the third month after delivery. Female eroticism reported in the postpartum period had no sure relationship to parity or age of the woman, but could be related directly to the act of nursing. Forty-seven of the 101 women (19 in the formerly nulliparous group, 28 in the parous group) described low or essentially negligible levels of sexuality during the reviews conducted early in the third postpartum month. A variety of reasons for lowered sexual tensions were presented. Excessive fatigue, weakness, pain with attempted coition, and irritative vaginal discharge were but a few of these reasons. The area of greatest expressed concern was personal fear of permanent physical harm if coital activity were resumed too soon after delivery.

The remainder of the group (21 formerly nulliparous and 33 parous women) reported varying levels of sexual interest. These reports ranged from rapid return to nonpregnant sexual tension levels within two to three weeks after delivery to significantly higher levels of tension than in the nonpregnant state, described as a group by the nursing mothers (24 women). Aside from constant concern by the entire group for the possibly injurious effects of too rapid return to coital activity, the only real deterrent to early sexual activity in this group was perineal pain or vaginal-barrel irritation after coition.

Interestingly, 11 women described significantly increased sexual pleasure derived from the tenderness of the episiotomy area or the tightness of the postpartum vaginal barrel. A major factor in the rapid return of eroticism among this group of women was the continued feeling of congestion and fullness in the pelvis that had been present in the second and third trimesters, which they associated with sex tension increment. The postpartum pelvis is in truth chronically congested with venous blood, so the subjective sensations described by these women are readily understandable.

Ten women from the previously nulliparous group and 14 from the parous group were successful in nursing their babies for at least two months after delivery. The highest level of postpartum sexual interest in the first three months after delivery was reported by this group of nursing mothers. Not only did they report sexual stimulation (frequently to plateau tension levels and, on three occasions, to orgasm) induced by suckling their infants, but as a group they also described interest in as rapid return as possible to active coition with their husbands. There was a heavy overlay of guilt expressed by 6 of the 24 women who admittedly were stimulated sexually by the suckling process. They were anxious to relieve concepts or fears of perverted sexual interest by reconstituting their normal marital relationships as quickly as possible. This concept has been reported previously [53, 263, 296] and is only confirmed by this investigation.

Of interest were the expressed concerns of 8 previously nulliparous and 17 parous women after voluntarily rejecting nursing opportunity during the postpartum period. A major factor leading to nursing refusal was verbal rejection of the concept of nursing by 16 husbands. Thirteen of these women expressed fear of loss of their figures and were not sufficiently reassured by medical authority to consider nursing in a positive vein. The remainder of the concerns ranged from personal rejection of the process as degrading to the fear expressed by 6 women of the high levels of eroticism stimulated by the suckling process. These 6 women were multiparas and had attempted nursing with previous pregnancies.

The fact that for at least four weeks after delivery all women are essentially castrates so far as ovarian function is concerned had no predictable influence on the reported levels of their eroticism or sexual performance. If negligible or absent ovarian-steroid production is assumed to play an all-important role in female sexuality, women in the immediate postpartum period would be expected to have low or absent levels of sexual interest until the ovaries resume sex-steroid secretion in normal physiologic sequence four to six weeks after delivery.

Obviously, as reported above, such was not the case. States of steroid starvation could have contributed to the excessive fatigue, exhaustion, and emotional instability that were reported as distressing symptoms among the low-sex-tension groups. However, the nursing women, normally contending with the longest delay in physiologic return of ovarian-steroid production, also had the highest levels of reported postpartum eroticism. There is no doubt that ovarian steroids have a role in female sexuality. That steroid levels play the all-important role, as has been presumed by so many in the past, obviously is false. Money and Lloyd have supported the concept of steroid fallibility [175, 232].

The incidence of masturbation during the postpartum period was essentially negligible, as only 5 women reported any recourse to or need for this form of relief. Two of these women were among the group that had delivered illegitimate pregnancies.

Fifty-eight of the 101 women reviewed during the postpartum period reported concern for their husbands' sexual tensions during the postpartum period of continence. Particularly were they concerned when postpartum continence was added to whatever pre-delivery-continence period had been established. Those women that had provided active relief for their husbands during third-trimester continence periods did so again after delivery. Three women in the parous and one in the previously nulliparous group had not approached their husbands before delivery, but during the postpartum period they assumed an active role in providing release for these men.

With the exception of the women for whom intercourse was interdicted medically for three months and the members of the nonmarried group, all women interviewed returned to full coital activity within six weeks to two months after delivery. Despite the fact that intercourse was prohibited for at least six weeks by most medical authority, there frequently was return to coital activity within three weeks of delivery by higher-tensioned women or by wives attempting to respond to male demand. Particularly was an early return to coition the pattern for those women actively nursing. The women that tended to follow physicians' prescribed periods of continence were those who had regained little of their own erotic interest.

MALE REACTIONS

As noted earlier in the chapter, 79 of the total of 92 available husbands cooperated with bisexual interrogation at the end of the third postpartum month, after all interrogative connection with their wives had been terminated. Their stated ages ranged from 25 to 49 years. Eight of the men reported previous marriages, and six had children by these marriages. Therefore, all experience could not be related to current or previous pregnancies by present wives. For the sake of brevity, only material relating to the last trimester of pregnancy and the postpartum period will be presented.

Thirty-one of the men stated in retrospect that they had withdrawn slowly, almost involuntarily, from active coital demand upon their wives toward the end of the second or early in the third trimester. As described earlier in the chapter, 20 of the 31 wives involved noted and reported their husbands' apparent loss of coital interest. The husbands gave no consistent reasons for withdrawal other than fear of causing physical injury to fetus or to wife. Eighteen men stated that they really weren't interested, and "don't know why." Only 5 of the men described the stigmata of pregnancy (swollen abdomen, thickened legs, etc.) as personally objectionable. Two men stated that their wives had not maintained personal cleanliness to the degree that was their pattern in the nonpregnant state and that they had lost sexual interest as a consequence.

Seventy-one of the men were married to women whose physicians had prohibited intercourse for periods varying from four weeks to three months prior to the estimated date of confinement. Twenty-one of these men stated that they understood, agreed with, and honored the prohibition. Twenty-three men did not understand the reason for the prescribed continence period, were not sure the doctor had said it, or wished that he had explained it to them as well as their wives. Twelve of the men reported that they had sought release outside of the home when denied conjugal opportunity by medical proscription.

The major concern expressed after delivery by the entire group was how soon active intercourse could be reinstituted without causing physical harm or emotional distress in their wives. None of the husbands had any clear idea of female postpartum physiology or psychology despite a liberal sprinkling of postgraduate degrees.

Several husbands of the high-tensioned postpartum women were concerned with an early return to sexual encounter that was established despite medical interdiction. They stated that the active demand for coition two to three weeks postpartum had been instituted by their wives.

The 12 men that sought release outside the marriage during the predelivery continence period continued to do so after delivery. Six more husbands described extramarital activity during the postpartum continence period. From 3 of these men came the declaration that this was the "first time."

CLINICAL CONCERNS

The study of female sexuality in pregnancy has highlighted material of basic obstetric relevance. This material is presented to suggest that little is known and frequently a great deal presumed in clinical areas relating to the pregnant woman's physiologic and psychologic patterns of sexual response.

For years, Javert [127, 128] has been concerned with the possible relation between female orgasmic experience during coition and pregnancy wastage in the first trimester. He has emphasized the importance of avoiding coital activity during the first trimester of pregnancy for individuals with a positive obstetric history of three successive abortions. However, there has been no similarly expressed concern for the severe uterine contractions that result from an orgasmic experience induced by manipulative activity during the first trimester of pregnancy.

Observations and physiologic recording of uterine contractile response to sexual stimuli in the nonpregnant state suggest that the uterus may contract with more intensity during an orgasmic experience accomplished by manipulative stimulation than from active coition (see Chapter 8). Similar investigation involving the six study subjects in the first trimester of pregnancy tends to support this clinical assumption (see Part 1 of Chapter 10). All six women subjectively described more severe uterine contractile response during an orgasm induced by manipulation than in orgasmic response to coital stimulation. If it is true that coitally induced orgasmic experience and the consequent uterine contractile response have a tendency to induce pregnancy wastage, particularly among susceptible women, it also must be considered highly probable that masturbatory activity will create the same tendency toward pregnancy wastage for the susceptible multiple aborter. Although both professional and lay literature [27, 95, 250, 295] has carried discussions of the problem of pregnancy wastage resulting from coition, warnings of the possibility of pregnancy wastage subsequent to successful erotic stimulation by manipulative techniques have not been given similar publicity.

Late in the third trimester of pregnancy, when the fetal presenting part enters the true pelvis and cervical effacement begins, there will be some reduction in the elevation of the anterior vaginal wall in response to sexual stimulation. This is purely a mechanical effect and does not reflect lack of expansive ability of the reacting tissues. With the presenting part deeply engaged in the true pelvis, the cervix is brought into the vaginal axis. At this stage, consistent, direct penile-cervical contact is probable during active coition. Some postcoital spotting may result from direct contact between the erect penile glans and the vasocongested predelivery cervix. If spotting or frank bleeding occurs, coition should be interdicted.

There are legitimate clinical concerns with both coition and automanipulation at or near term. It probably is true that the contractions of orgasm at or near term can send a woman into labor. Certainly this reaction has been verbalized sufficiently to be of some possible clinical significance. Four women, uninvolved in the research program, have reported the onset of labor immediately subsequent to orgasmic experience [131]. In three instances, orgasmic response was initiated by coital stimulation. In one situation, the contractions initiated by multiorgasmic automanipulation continued into full-fledged labor with delivery. In all four cases the women were within 18 or less days of the expected date of confinement. Three women were having a first and one a second delivery. No obstetric distress was encountered. Whether or not premature labor can or has been induced by orgasmic response is of major clinical moment. There is no secure information available on this subject.

Insofar as the clinical problem of infection resultant from coition is concerned, this problem applies not only to the last few weeks but also to any stage of pregnancy, or for that matter, to the nonpregnant state. The interdiction of coition during the latter part of the third trimester based only upon fear of infection for mother or child is a residual of the preantibiotic days in medicine and largely can be negated. Infection of the vaginal barrel immediately prior to labor certainly is controllable medically. In like manner, full protection can be and is provided the newborn infant.

The uterus remains an abdominal organ for approximately six weeks after delivery. Bleeding from the placental site on the wall of the uterus usually continues for two to four weeks postpartum. Incisions in the perineum and posterior vaginal wall (episiotomy) made to aid in the delivery of the child usually are well healed within two to three weeks. Therefore, from a purely physiologic point of view, there is no contraindication to coition once the postpartum vaginal bleeding has stopped and any incisions or tears in the vaginal outlet have healed.

The whole problem of coition during the third trimester of pregnancy and the postpartum period should be individualized. As described, late in the third trimester there is possible direct penile contact with a relatively unyielding cervix when the presenting fetal part is deeply engaged in the pelvis. Yet many women, particularly multiparas, go through the entire third trimester without deeply engaging the presenting part until in active labor. There is no real contraindication to coition in this situation insofar as damage to the pregnancy on a purely mechanical basis is concerned.

Although the female may be physiologically capable of resuming coition early in the postpartum period, she may or may not be psychologically ready to do so. Again the problem should be considered by medical authority on an individual basis. Many women are anxious to return to coital connection as soon as physically possible. They should be encouraged in this regard. There are those women who prefer longer periods of continence. Their situation should be discussed, personal reasons examined, fears explained away, and a firm understanding between both members of the marital unit established. Frequently, blanket medical interdiction of coital activity for arbitrarily established periods of time both before and after delivery has done far more harm than good.

A more thorough investigation of problems of sexual activity for both husband and wife late in the third trimester and early in the postpartum period is in order. Physicians often fail to take into account the factor of long-term male continence. Six weeks before and six weeks after delivery usually are proclaimed restricted periods by medical interdiction. Many male partners first break marital vows during this three-month period. If such a degree of sexual continence is in order, the situation should be explained to the marital partners, and concerns with automanipulation also should be discussed. If continence is not indicated, a healthy marriage may be preserved by individualizing each case. Many men accustomed to regular ejaculatory release find three months an excessive period of continence and may elect not to tolerate such degrees of personal restraint.

The material presented in this chapter reflects the influence of pregnancy upon female eroticism. Although the sample obviously is prejudiced these reports suggest that parity and trimester of pregnancy exert the greatest influence upon female sexual response before delivery, and that nursing in the postpartum period usually initiates rapid return of female interest in sexual performance. Male concern with the relation of a state of pregnancy to marital-unit sexuality has been approached casually and needs further amplification. Possibly of significance is the suggestion that medical authority should individualize rather than arbitrarily assign prepartum and postpartum periods of continence.

MALE SEXUAL
RESPONSE

I I

MALE EXTRAGENITAL REACTIONS

The human male's physical reaction to elevated levels of sexual tension is not confined to the primary or secondary organs of reproduction. Physical evidence of sexual tension develops throughout the entire body. Just as with the female, physiologic response to effective sexual stimulation follows two basic patterns—first, widespread vasocongestion (superficial and deep) and second, myotonia (generalized and specific). Physical reactions other than those involving the organs of reproduction are of sufficient magnitude to merit separate consideration.

THE BREASTS

There is no consistent anatomic response of the male breast to sexual stimulation. However, frequent nipple erection and additional nipple tumescence have been demonstrated by male study subjects. If nipple erection is to occur, it usually develops late in the excitement phase and lasts throughout the remainder of the sexual cycle. Tumescence of the erect nipples is evidence of plateau-phase levels of sexual stimulation. Nipple erection has been observed in 60 percent of the study subjects and additional tumescence in 7 of 10 males who attain full nipple erection. However, late excitement or early plateau vasocongestive reactions of increased venous pattern or areolar tumescence have not been observed in the sexually responding male.

There are no specific male breast reactions during the late plateau or orgasmic phases of the sexual response cycle. If both primary erection and secondary tumescence of the nipples are established, it may be many minutes or even more than an hour after ejaculation before resolution-phase nipple retraction is accomplished.

Nipple erection and tumescence usually develop without direct contact. As a source of erotic stimulation male breasts and nipples seldom are manipulated directly during heterosexual activity. However, breast stimulation does constitute a significant segment of male homosexual activity. As a result, the nipples and even the anterior chest wall develop erotogenic qualities seldom found in the heterosexually oriented male.

SEX FLUSH

The human male develops the same maculopapular sex flush described for the human female (see Chapter 3). A statement of statistical rate of appearance will be omitted, however, since the flush occurs with wide variation within the same individual as well as between different individuals. Although an overall observation of 25 percent incidence was noted throughout the study, correlation with individual tendency or individual circumstance did not seem sufficiently pertinent to record.

When the sex flush occurs in the male, it arises from the epigastrium and spreads over the anterior chest wall. The neck, face, and forehead progressively are involved. Occasional evidence of the flush appears on the shoulders, forearms, and thighs of particularly susceptible males. When fully developed, the maculopapular sex flush resembles the measles rash. While this flush may appear late in the excitement phase in a situation of rapid tension elevation, it usually develops after the plateau phase is well established. Thereafter it spreads rapidly as the highly stimulated male progresses toward an orgasmic experience.

The appearance of this measles-like rash in any phase of the cycle provides evidence of high levels of sexual tension. For example, a man may experience a complete cycle of sexual response without manifesting a sex flush. During a subsequent opportunity much higher levels of sexual tension may develop and the sex flush may become quite evident. The actual occurrence and timing of onset of the male sex flush seem to be influenced by both environmental and psychogenic factors. Evidence of this is suggested when recorded incidence is higher in a heated rather than a cool room and in situations of extreme anticipation rather than routine performance.

During the refractory period of the male's resolution phase the sex flush disappears with extreme rapidity. It disappears initially from the shoulders and extremities, secondarily from the diaphragm and anterior chest wall and, finally, from the neck, face, and forehead.

MYOTONIA

Myotonia becomes clinically obvious late in excitement-phase and during plateau-phase levels of sexual tension and is both generalized and specific in character. Usually muscles contract with regularity or in spasm in an involuntary manner, but contraction frequently may be voluntary, depending upon coital positioning.

Carpopedal spasm [144, 203, 239] rarely has been observed with the male in the usual superior coital position (see Chapter 18). The physical activity associated with this position necessitates employment of voluntary musculature of the trunk, pelvis, and extremities and usually precludes development of involuntary striated-muscle spasm in the extremities. If the male is in supine position during coition, carpopedal spasm occurs frequently. This involuntary spasm of the striated musculature of the hands and feet is an indication of high levels of sexual tension. Carpopedal spasm has been observed more frequently during male masturbatory episodes than during intercourse, regardless of body positioning.

A detailed discussion of both male and female myotonia during sexual response has been undertaken in Chapter 18.

THE RECTUM

The external rectal sphincter contracts irregularly subsequent to direct stimulation in both the excitement and plateau phases of the sexual response cycle. During the ejaculatory experience the sphincter also contracts involuntarily. Orgasmic contractions of the sphincter recur in regular patterns with an 0.8-second intercontractile interval. They occur simultaneously with the expulsive contractions of the periurethral musculature, but do not recur more than two to four times. Resolution-phase relaxation of external-sphincter muscle tension occurs before the expulsive contractions of the penile urethra have been completed.

CARDIORESPIRATORY REACTIONS

HYPERVENTILATION

Hyperventilation developing during the late plateau and the orgasmic phases of the sexual response cycle is a normal occurrence. The physiologic intensity and the duration of the hyperventilative reaction frequently are indicative of the degree of sexual tension. The reaction usually continues through the orgasmic phase and resolves in the refractory period of the resolution phase. Respiratory rates peaking at over 40 per minute have been recorded repeatedly during orgasm.

TACHYCARDIA

The heart rate of the responding male increases in direct parallel to his rising sexual tensions. During plateau-phase levels of sexual tension the heart rates of male study subjects have been recorded in a range of 100–175 beats per minute. The slower the initial heart rate at resting stage, the lower the rate during sexual stimulation. Orgasmic-phase recordings have ranged from 110 to 180+ beats per minute. Electrocardiograph tracings must be studied with a magnifying glass to count the beats per minute at the extremely high levels of cardiac contractility. These rates occur regardless of the technique of sexual stimulation (Fig. 11-1).

BLOOD PRESSURE

Elevations in systolic blood pressure ranging from 40 to 100 mm. Hg. have been recorded from male study subjects during coition and

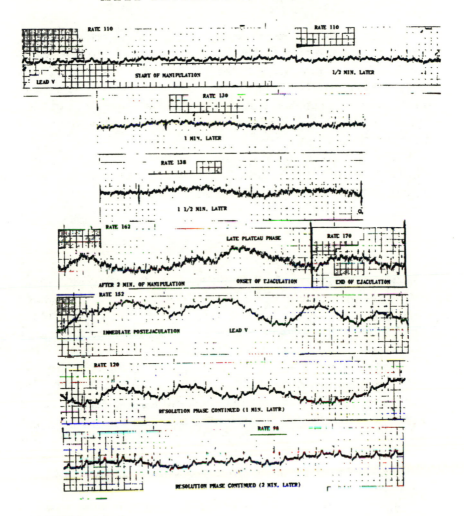

FIGURE 11-1

Male study-subject, electrocardiogram (Lead V): cardiac rates during sexual stimulation.

automanipulation. Diastolic pressure elevations have ranged from 20 to 50 mm. Hg.

In the past there have been several attempts to evaluate the problems inherent in the cardiorespiratory response to sex tension increment [22, 67, 151, 152, 220, 282]. Clinicians daily face the problem of advising the patient recovering from an acute episode of coronary artery occlusion. One of the immediate problems is that of the family unit's interest in returning to an active sexual relationship. How much cardiac strain develops in response to sexual tensions? Is coition more of a strain on the cardiac patient than is masturbation? Are there sex techniques that will place less strain on the heart, yet relieve the individual and the family unit of their tension increment? These are but a few of the questions that medicine must answer.

A report on male cardiorespiratory response to sexual tensions will be published in the future, with detailed consideration given to both general body and target-organ physiology.

PERSPIRATORY REACTION

Many males evidence an involuntary sweating reaction immediately after ejaculation. This perspiratory reaction may develop whether or not there has been obvious physical exertion during the sexual encounter, and whether or not a sex flush has appeared. Usually this perspiratory reaction is confined to the soles of the feet and the palms of the hands, but may appear on the trunk and occasionally may involve the head, face, and neck of the responding male. If perspiration appears on body surfaces, it does so during the refractory period of the resolution phase. Occasionally the perspiratory reaction develops so rapidly that its appearance is concomitant with the final ejaculatory contractions of the penis. Approximately one-third of all male study subjects developed the perspiratory reaction.

Human physiologic response to sexual stimuli is above all else a protean reaction. The examples of reaction in specific body areas or organ systems recorded in this chapter should not distract from the concept that the entire body is involved by sex tension increment. As the tensions elevate, so do the reactions of vasocongestion and myotonia, until the trigger point of physiologic release, orgasm, is attained.

I 2

THE PENIS

1 . ANATOMY AND PHYSIOLOGY

The anatomy and physiology of penile-erection response to sexual stimulation have been considered exhaustively in the literature [19, 94, 142, 276, 299, 329]. This material will be presented in detail sufficient only to establish orientation to the male's primary reproductive viscera.

The morphology of the penis provides ideal support for the primary physiologic response to sexual stimulation, vasocongestion. The penis is formed of three cylindrical bodies of erectile tissue (Fig. 12-1). Two of the cavernous cylinders, the corpora cavernosa penis, lie parallel to each other and just above a third cylinder of erectile tissue, the corpus spongiosum, which, in addition to erectile tissue, contains the urethra. The corpus spongiosum is expanded at its base to form the urethral bulb and distally to form the glans penis. The two corpora cavernosa and the corpus spongiosum are each surrounded by an individual fibrous coat, the tunica albuginea, and all three corpora are enclosed in dense fascial capsules.

At the base or root of the penis the corpora cavernosa penis diverge to form the crura, two processes which are attached directly to rami of the pubis and ischium (the pubic arch). Each of the crura is sheathed by the ischiocavernosus, a skeletal muscle. The corpus spongiosum also is encapsulated by a skeletal muscle, the bulbospongiosus muscle.

The two corpora cavernosa and the corpus spongiosum form the erectile tissue of the penis, receiving arterial blood from branches of the internal pudendal arteries. These branches are (1) the dorsal arteries of the penis found near the dorsal surface of the penis in the tunica albuginea; (2) the cavernous arteries running longitudinally through each corpus cavernosum penis; and (3) two bulbourethral arteries that run longitudinally through the corpus spon-

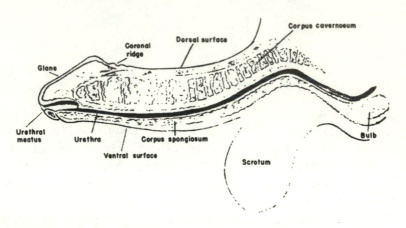

FIGURE 12-1

The penis: normal anatomy (lateral view).

giosum just ventral to the urethra. Terminal branches from these arteries, the helicine arteries, end in small capillaries that open directly into the cavernous spaces (Fig. 12-2).

Venous return is by two pathways, the superficial dorsal vein which drains the entire corpus spongiosum including the glans and the urethral bulb, and the deep dorsal vein that drains the corpora cavernosa.

The cavities of the three corpora cavernosa serve as erectile tissue. There are many compartments separated by bands or cords of fibrous and smooth-muscle tissue called trabeculae. These compartments are interspersed with arterioles the intima of which sup-

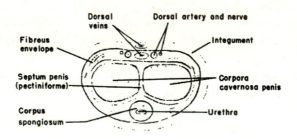

FIGURE 12-2

The penis: normal anatomy (transverse view).

posedly contains ridges which in a contracted state act to constrict partially the blood flow entering the cavernous sinuses. When the arterioles are dilated, the flow of blood into the penis is increased and the sinuses are filled. The veins of the penis are believed to possess valves that slow down the return of blood from the penis [87, 148]. Previously, contraction of the ischiocavernosus muscle was believed to aid in erection by secondary venous constriction, but little support is given now to this concept [108].

The return of the erect penis to a flaccid state probably is much more the result of active constriction of the arterioles than of any other action. The trapped blood then escapes from the cavernous sinuses despite any existent passive constriction of the veins.

Dilatation of the penile arteries and subsequent penile erection are the result of stimulation of the splanchnic nerves. Erection is lost when the sympathetic nerve supply causes constriction of the penile arteries. A center for reflex erection is believed to exist in the sacral portion of the spinal cord [110, 285]. Obviously, stimulation of erection primarily is directed from the higher cortical centers [10, 25, 238, 307, 320].

The erotic component of the male genitalia has been concentrated in external pelvic anatomy. The penis, scrotum, and rectum are markedly sensitive to sexual stimulation, while the prostate, seminal vesicles, vas deferens, etc., have little or no sensual focus (Fig. 12-3). These structures respond to effective sexual stimulation both by vasocongestion and by elevated muscle tension. It should not be presumed, however, that every incidence of increased vasocongestion or of muscle tension evidenced by male genitalia necessarily reflects states of elevated sexual tension. While recurrent erection and detumescence of the penis usually are evidence of waxing and waning sexual tensions, there are other causes for penile erective response that should be considered in context before attention is focused specifically on penile reaction to sexual stimulation. The penile erection that adolescent and mature males frequently evidence upon awakening has been observed on multiple occasions. Partial penile erection subsequent to stress on the perineal musculature, such as that resulting from lifting unusually heavy loads or straining at stool, has been recorded. Involuntary penile erection

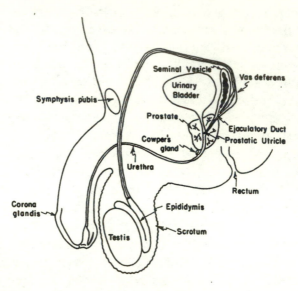

FIGURE 12-3

Male pelvis: normal anatomy (lateral view).

also has been demonstrated repeatedly clinically in such pathologic conditions as phimosis with irritative proprioceptive stimulation of the glans, and benign prostatic hypertrophy with secondary penile stimulation resultant from chronic urinary retention.

Recently, regularly recurrent penile erection has been recorded during the observed sleep of male experimental subjects [131]. In younger males dream erections recur regularly despite ejaculatory experience in an immediate presleep sequence. However, if the male is in his late thirties or beyond, an ejaculatory experience immediately before observed sleep reduces the incidence of dream erections, particularly in the first few hours of the sleep sequence.

Hyperinvolution of the penis beyond resolution-phase levels of detumescence has been observed clinically on numerous occasions. Penile involution following exposure to cold (e.g., swimming in cold water) is well established. In situations of acute exhaustion consequent to severe physical strain, the penis usually is smaller than its normal flaccid size. Advancing age or surgical castration may and frequently does produce a secondary involution of the penis which permanently reduces organ size below previously established normal states for the individual involved.

Unpublished data also suggest that the penis of the secondarily impotent male attains states of pathologic hyperinvolution (when compared to previously established norms), after two to four years of unremitting impotence [131]. Particularly does hyperinvolution become clinically obvious immediately following attempted and failed sexual encounter. Clinical observations tend to support the possibility that penile hyperinvolution, like penile erection, although frequently developed on a reflex basis, may also respond directly to higher cortical centers.

Since neither the physiology of penile hyperinvolution nor of penile response to asexual stimuli has been investigated, these problems cannot be explored at this time. A consideration of penile response to sexual stimuli will provide a baseline upon which future studies of these reactions can be based.

The initial physiologic response of the human male to effective sexual stimulation is erection of the penis. This reaction is the neurophysiologic parallel to the human female's production of vaginal lubrication [202]. Both reactions occur with equal facility, regardless of whether the source of sexual stimulation is initially somatogenic or psychogenic in origin (see Chapter 17).

During eleven years of direct investigation of human sexual response, penile erection has been observed in males of all ages, ranging from baby boys immediately after delivery to men in their late 80's. Penile erection in the neonate is considered to be evidence of specific vasocongestion and of increased neurogenic and myogenic irritability as might be stimulated by a severe crying spell. Penile reactions of aged men usually are so varied that descriptions of response patterns have been individualized (see Part 1 of Chapter 16). The mere fact that full penile erection could be obtained and an ejaculation produced by an 89-year-old man during episodes of active cooperation as a study subject is considered worthy of report.

In order to facilitate clinical consideration of penile response to sexual tensions, the cycle of sexual response again will serve as a framework upon which descriptions of anatomic and physiologic reaction can be placed in proper continuity.

EXCITEMENT PHASE

The human male's first physiologic response to effective sexual stimulation is penile erection (Fig. 12-4). There may be only a minimum degree of sexual tension present before this response pattern has been completed. After full penile erection has been attained, the excitement phase may extend for the briefest of intervals or for a matter of many minutes in direct parallel to the intensity of or variation in any form of successful sexual stimulation.

Penile erection has been maintained by study subjects for extended periods by carefully controlling variation and intensity of stimulative techniques. Erection has been partially lost and subsequently rapidly regained many times during an intentionally prolonged excitement phase. With variation from intense somatogenic stimulation to complete recession of such activity, penile tumescence may increase or decrease repeatedly over long periods of time without either the achievement of full penile vasodistention or total loss of vasocongestion.

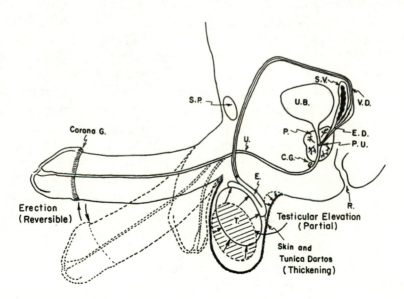

FIGURE 12-4

Male pelvis: excitement phase.

Psychosensory diversion has been created frequently in the laboratory during excitement-phase response. Penile erection may be impaired easily by the introduction of asexual stimuli, even though sexual stimulation is continued simultaneously. Despite constantly maintained somatogenic penile stimulation, a sudden loud noise, vocalization on an extraneous subject, or an obvious change in lighting, temperature, or attendant personnel may result in partial or even complete loss of penile erection.

PLATEAU PHASE

The penis that apparently has achieved full erection during excitement phase undergoes a minor involuntary vasocongestive increase in diameter as the orgasmic (ejaculatory) phase approaches (Fig. 12-5). This additional plateau-phase tumescence is confined primarily to the corona glandis area of the glans penis.

A color change also may develop in the glans penis late in the plateau phase of the sexual cycle. There may be a deepening of the

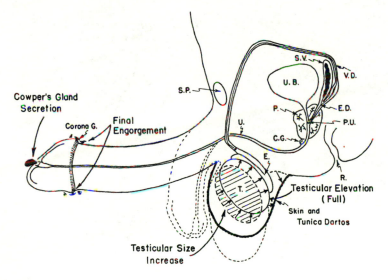

FIGURE 12-5

Male pelvis: plateau phase.

mottled reddish-purple color of venous stasis. This preejaculatory color change, when well established, is reminiscent of the preorgasmic discoloration of the minor labia of the human female. However, the color change of the corona glandis is not a constant occurrence, nor as well defined, when it does develop, as the minor-labial sex-skin reaction of the human female (see Chapter 4). Approximately 20 percent of all male study subjects have demonstrated the coronal color change. It should be emphasized, however, that some men may develop a coronal color change during one cycle and not during a subsequent sexual encounter. The appearance of the coronal color change has been too variable to allow objective conclusions to be drawn as to its relation to severity of response or the influence of such variables as type of stimulation, length of continence, etc., on its appearance or intensity.

ORGASMIC PHASE

The orgasmic-phase penile ejaculatory reaction (Fig. 12-6) develops from regularly recurring contractions of the sphincter ure-

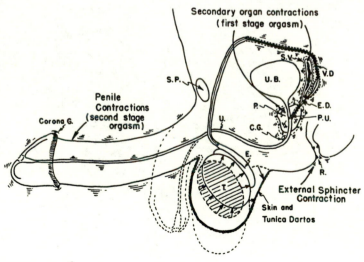

FIGURE 12-6
Male pelvis: orgasmic phase.

thrae, bulbospongiosus, ischiocavernosus, and transverse superficial and deep perineal muscles. The ejaculatory contractions involve the entire length of the penile urethra and force the seminal-fluid content from the prostatic and membranous portions of the urethra to and through the urethral meatus. The seminal fluid is expelled the full length of the penile urethra under severe pressure created by the involuntary but coordinated contractions of these muscle groupings.

The intercontractile intervals of the initial penile contractions have been timed at a rate similar to that of the orgasmic-platform contractions which develop in the vagina during the female's orgasmic experience. Expulsive penile contractions start at intervals of 0.8 second. After the first three or four major expulsive efforts the penile contractions are rapidly reduced in frequency of recurrence and in expulsive force. Minor contractions of the penile urethra continue for several seconds in an irregularly recurrent manner, projecting a minimal amount of seminal fluid under little if any expulsive force. The terminal intercontractile intervals are extended to several seconds in duration.

RESOLUTION PHASE

Resolution-phase penile detumescence develops in two distinct stages (Fig. 12-7). The primary stage of penile involution, occurring early in the refractory period of the resolution phase, reduces the penis from full erection to approximately 50 percent larger than its unstimulated, flaccid state. This primary stage of penile detumescence usually occurs with extreme rapidity. Secondary-stage penile involution, which ultimately returns the penis to its normal, unstimulated size, may be an extended involutionary process which lasts well past the refractory period of the resolution phase.

The primary stage of penile involution usually is prolonged when the excitement or plateau phases of the particular sexual cycle have been extended by direct intent. Many males learn to restrain or delay their ejaculatory reaction until their sexual partner is satiated. Satiation on the woman's part may represent several complete cycles of sexual response with the consequent demand for maintained

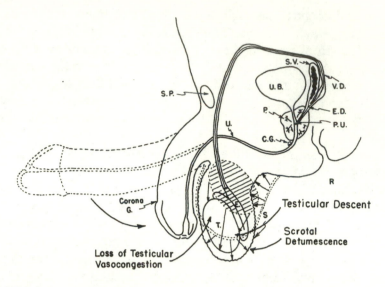

FIGURE 12-7

Male pelvis: resolution phase.

penile erection for extended periods of time. When an erection is maintained for many minutes, particularly during long-continued intravaginal containment, advanced degrees of penile vasocongestion frequently continue after the actual ejaculatory experience. Thus, the primary stage of penile involution, usually a rapid process, may be extended indefinitely and second-stage penile involution subsequently delayed. No acceptable physiologic explanation can be offered at present for this clinical observation.

The rapidity with which total penile (two-stage) involution is accomplished frequently is dependent on the existence and intensity of extraneous stimuli. The detumescent process is slowed during the second stage when residual sexual stimulation persists, and is speeded when the stimuli are asexual in character. If the penis is removed from the vagina immediately following an ejaculatory experience, full detumescence is accomplished much more rapidly than if the postejaculatory penis remains in the stimulative vaginal barrel. If the male simply maintains close physical proximity to his female sexual partner, the secondary stage of penile detumescence may be quite prolonged. With opportunity for long-continued bodily contact with a sexual partner, the penis may not complete a second-stage detumescence and revert to its normal flaccid state for many minutes after an ejaculatory experience.

If the resolution-phase male walks about, talks on any extraneous subject, or is otherwise diverted in an asexual manner, secondary-stage penile involution occurs with relative rapidity. For example, if the male attempts to urinate immediately after an ejaculatory experience, the involuntary mental concentration directed toward the urinary process will shorten both primary and secondary stages of penile detumescence. Actual urinary effort always will increase the rate of penile detumescence, since the male cannot urinate with the penis in full erection. If the penis is still somewhat enlarged (secondary-stage involution) at the onset of micturition, the penis usually is in a completely flaccid state by the time the act is completed.

Finally, the physiologic response of the penile urethra to sexual tensions should be described. Obviously, the urethra lengthens during active penile erection. As excitement-phase progresses toward plateau, the lumen of the penile urethra undergoes at least a twofold increase in transverse diameter. As the plateau phase of sexual tension is experienced, this increase in penile urethral diameter approaches threefold magnitude at the base of the urethra where the urethral bulb is located (Fig. 12-8). Late in the plateau phase there

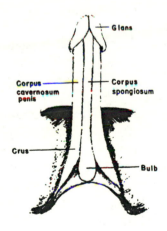

FIGURE 12-8

The penis: erect (ventral view).

is further transitory distention of the urethral bulb that varies in degree from man to man. This preorgasmic rapid distention of the urethral bulb is pathognomonic of impending orgasm and is a reaction of such magnitude as to be obvious to direct observation.

The penile urethra contracts in rhythm parallel to that of the total penile body during the ejaculatory process.

The urethral meatus usually becomes slightly patulous in its transverse axis during the excitement phase of the male's sexual response cycle. There is no further plateau- or orgasmic-phase dilation of the meatus. Direct observation of the urethral meatus provides no clue to the imminence of an ejaculatory experience. The urethral meatus loses its patulous transverse distention immediately after ejaculation.

During the refractory period of the resolution phase loss of increased urethral diameter and shrinkage of the overdistended urethral bulb occur before loss of increased urethral length can be determined.

2. CLINICAL CONSIDERATIONS

The functioning role of the penis is as well established as that of any other organ in the body. Ironically, there is no organ about which more misinformation has been perpetrated. The penis constantly has been viewed but rarely seen. The organ has been venerated, reviled, and misrepresented with intent in art, literature, and legend through the centuries. These intentional misrepresentations have varied in magnitude with the culture. Our culture has been influenced by and has contributed to manifold misconceptions of the functional role of the penis. These "phallic fallacies" have colored our arts and, possibly of even more import to our culture, influenced our behavioral and biologic sciences.

The twofold functioning role of the penis providing for both urinary release and seminal-fluid deposition has been accepted throughout recorded history. Why, with the functioning role unquestionably established, should the functional role of the penis have been shrouded so successfully by "phallic fallacy" concepts? This, indeed, is one of the great mysteries of biologic science.

The functional role of the penis is that of providing an organic means for physiologic and psychologic increment and release of both male and female sexual tensions. The penis as an organ of male sensual focus can be related to the functional role of the clitoris (see Part 2 of Chapter 5) in the total of female sexual response. The gross difference between these two organs is that the clitoris serves only in a functional role, and the penis has both a functioning and a functional capacity.

Objective material relating to the functional role of the penis has been accumulated from over 2,500 directly observed sexual response cycles experienced by 312 male study subjects whose ages range from 21 to 89 years. Subjective material has been returned from team interrogation of 654 men screened as study-subject applicants before the 312 active participants were selected. Material for anatomic consideration has been developed from direct evaluation of penile and scrotal content plus rectal examination for prostate and seminal-vesicle anatomy. These examinations were conducted routinely on all males who became active members of the study-subject population.

Additional material of both physiologic and psychologic content has been returned from investigation of human sexual inadequacy. This clinical-research program has been running concurrently with the basic-science investigation of human sexual response for the past seven years.

CIRCUMCISION

Thirty-five of the 312 male members of the study-subject population were uncircumcised. Although approximately one-quarter of the male study subjects were beyond 40 years of age, more than half (19) of the uncircumcised males were found in this age grouping. The fact that only 16 out of a total of 231 male members of the study-subject population between the ages of 21 and 40 years were uncircumcised is representative of the medical trend toward urging routine circumcision of the newborn male infant. More than

95 percent of all deliveries in this country now are hospital deliveries, and circumcision is recommended as a routine neonatal procedure. The uncircumcised male, particularly one born in an urban area, indeed, is becoming a rarity in our society.

The phallic fallacy that the uncircumcised male can establish ejaculatory control more effectively than his circumcised counterpart was accepted almost universally as biologic fact by both circumcised and uncircumcised male study subjects. This concept was founded upon the widespread misconception that the circumcised penile glans is more sensitive to the exteroceptive stimuli of coition or masturbation than is the glans protected by a residual foreskin. Therefore, the circumcised male has been presumed to have more difficulty with ejaculatory control and (as many study subjects believed) a greater tendency toward impotence.

A limited number of the male study-subject population was exposed to a brief clinical experiment designed to disprove the false premise of excessive sensitivity of the circumcised glans. The 35 uncircumcised males were matched at random with circumcised study subjects of similar ages. Routine neurologic testing for both exteroceptive and light tactile discrimination were conducted on the ventral and dorsal surfaces of the penile body, with particular attention directed toward the glans. No clinically significant difference could be established between the circumcised and the uncircumcised glans during these examinations.

A clinical observation specifically relating to the uncircumcised penis may explain in part this lack of excessive sensitivity of the circumcised as opposed to the uncircumcised glans. Frequently during coition and occasionally during automanipulation the foreskin of the uncircumcised male retracts from the glans as the fully erect penis reacts to plateau-phase levels of sexual tensions. The foreskin retracts in direct relation to the degree of freedom of its movement over the subjacent glans with the penis in a flaccid state. When a minor to moderate degree of phimosis is present, foreskin retraction occurs only after long-continued coital connection. Only 6 of the 35 uncircumcised study subjects failed to demonstrate significant exposure of the glans during or immediately subsequent to active coition. Foreskin retraction usually does not develop as frequently, or progress as far, during automanipulation as during coition. This fact probably is related to the manipulative techniques employed. These will be discussed later in the chapter.

Since 29 of the 35 uncircumcised males developed a significant degree of foreskin retraction during active coition, obviously the uncircumcised glans frequently is exposed directly to exteroceptive stimuli resultant from intravaginal containment just as is the circumcised glans. Thus from a physiologic point of view, a retained foreskin probably contributes little if anything to the individual male's ejaculatory control.

PENILE FALLACIES

Another widely accepted "phallic fallacy" is the concept that the larger the penis the more effective the male as a partner in coital connection. The size of the male organ both in flaccid and erect state has been presumed by many cultures to reflect directly the sexual prowess of the individual male. Dickinson [56] was one of the first to record dimensions of the penis with some degree of objectivity. He supported Loeb's report [176] that the normal range of penile length varies from 8.5 to 10.5 cm. in the flaccid state, with the general average in the 9.5 cm. range. The range of normalcy suggested by these measurements also has been supported by measurements returned from examinations of individual members of the male study-subject population [131].

The delusion that penile size is related to sexual adequacy has been founded in turn upon yet another phallic misconception. It has been presumed that full erection of the larger penis provides a significantly greater penile size increase than does erection of the smaller penis. This premise has been refuted by a small group of men selected from the study-subject population for clinical evaluation. Forty men whose penises measured 7.5–9 cm. in length in the flaccid state were compared to a similar number of study subjects whose penises in the flaccid state measured 10–11.5 cm. Measurement was crudely clinical at best and can only be presumed suggestive and certainly not specific in character. The length of the smaller penises increased by an average of 7.5–8 cm. at full (plateau-phase) erection. This full erection essentially doubled the smaller organs in length over flaccid-size standards. In contrast, in the men whose organs were significantly larger in a flaccid state (10–11.5 cm.), penile length increased by an average of 7–7.5 cm. in the fully erect (plateau-phase) state.

These measurements of full penile erection are so crudely clinical that they have been adjusted arbitrarily to the nearest 0.5 cm. to facilitate presentation. In each instance, measurement was taken from the anterior border of the symphysis at the base of the penis along the dorsal surface to the distal tip of the glans. All 80 penises were measured on three different occasions both in flaccid and erect states by the same individual. Only one investigator conducted this clinical measurement so that any idiosyncrasy of measurement technique would be common to all results. One of the measurements of penile erection was taken during automanipulation, and two measurements were initiated immediately upon withdrawal of the plateau-phase penis from active coition. Measurement of an erect penis was not attempted until the final engorgement of late plateau phase had been accomplished. Since full penile engorgement is a short-term process before ejaculation intervenes, measurement frequently was rushed and, therefore, additionally unreliable. While the information returned obviously is not definitive, there certainly is no statistical support for the "phallic fallacy" that the larger penis increases in size with full erection to a significantly greater degree than does the smaller penis. The difference in average erective size increase between the smaller flaccid penis and the larger flaccid penis is not significant.

Of clinical interest is the fact that the greatest observed penile-size increase from flaccid to erect state occurred in a male study subject with an average flaccid measurement of 7.5 cm. (not included in the experiment reported above). The increase in size from flaccid to erect state was just over 9 cm. This penis more than doubled in length when reacting from flaccid to erect state. The smallest increase in size from flaccid to erect state was observed in one of the larger organs. This penis measured just under 11 cm. in its flaccid state, yet at full erection only 5.5 cm. had been added to the length of this larger organ. At full plateau-phase erection the two organs were measured at identical lengths on three separate occasions.

As Piersol has stated [248], the size of the penis has less constant relation to general physical development than that of any other organ of the body. This statement has been made in recognition of yet another "phallic fallacy." It has been presumed by many cultures that the bigger the man in skeletal and muscular development, the bigger the penis, not only in a flaccid but also in an erect state. Detailed examination of the study-subject population of 312 men aged 21 to 89 years supported Piersol's contention that there is no relation between man's skeletal framework and the size of his external genitalia. The largest penis in the study-subject population, measuring approximately 14 cm. long in the flaccid state, was in a man 5 feet, 7 inches tall weighing 152 pounds. The smallest penis, measuring just over 6 cm. in the flaccid state, was in a man 5 feet, 11 inches tall weighing 178 pounds.

Although there is little to support the concept that erective size is proportionally greater for the larger than the smaller penis, there remains the theoretical concern of the man with the small penis as to his potential coital effectiveness. Even with erective ability of the smaller penis (less than 9 cm.) presumed equal to that of the larger penis (more than 10 cm.), the smaller penis in the flaccid state usually remains somewhat smaller in an erect state. The factor that constantly is overlooked in theoretical discussions of penile coital effectiveness is the involuntary accommodative reactions of the vagina in its functional role under coital stimulation as a seminal receptacle (see Part 2 of Chapter 6).

VAGINAL FALLACIES

The vagina is infinitely distensible from a clinical point of view. If mounting occurs early in the excitement phase, before involuntary expansion in length and transcervical diameter has developed fully, the woman may experience immediate difficulty in accommodating an erect penis, particularly a large organ. Presuming sufficient physical distress does not occur with intromission to dispel the woman's sexual tensions, involuntary vaginal expansion continues rapidly. Full accommodation usually is accomplished with the first few thrusts of the penis, regardless of penile size. If intromission occurs early in the woman's sexual response cycle, the fully erect smaller penis can and does function as a dilating agent as effectively as a larger penis.

With advanced excitement or early in plateau phase, the vagina normally overextends in length and overexpands at transcervical depth (see Part 1 of Chapter 6). This elliptical vaginal expansion, creating an anatomic basin for the imminent seminal pool, accounts for some loss of exteroceptive stimulation of the distal half of the fully inserted penis and reduces vaginal sensate focus for the female. Before the orgasmic platform in the outer third of the vagina develops sufficiently to provide increased exteroceptive and proprioceptive stimulation for both sexes, the overdistended excitement-phase vagina gives many women the sensation that the fully erect penis (regardless of size) is "lost in the vagina."

The obstetrically traumatized vaginal barrel increases accommodative difficulties for many women. Some vaginas are perpetually enlarged as the result of tears of pelvic fascia and musculature caused by childbirth, resulting in secondary cystoceles, rectoceles, and chronic cul-de-sac distention. In addition to sustaining the fascial and muscle tears these women also may have lost muscle tone throughout the pelvic area, as Kegel has emphasized repeatedly [138, 139]. These traumatized vaginas so overexpand late in excitement or during plateau phases that the resultant enveloping rather than supporting and constricting vaginal barrels inevitably reduce susceptibility to exteroceptive and proprioceptive stimuli for copulating women much more than for their male partners. This local reduction in sensate focus is more than counterbalanced by the psychologic stimuli initiated by the act of copulation (see Chapter 9).

Female concept of vaginal anatomy often provides a female counterpart to the fallacies occasioned by male anatomic misconceptions. There are occasional women with exceptionally large or small vaginas, just as there are occasional men with an exceptionally large or small penis [131]. The large vagina reacts as an obstetrically traumatized vagina and expands involuntarily far beyond the point of physiologic demand. Only one exceptionally large vagina was identified among the study-subject population of 382 women. This anatomic anomaly rendered immediate accommodation regardless of the size of the artificial penis introduced. A normal-sized penis could be accommodated in this large vagina without an obvious reaction. Therefore, there was little involuntary distention or "tenting" of the vaginal walls. With a twice-normal-sized penis introduced, the large vagina evidenced the involuntary accommodation reaction, expanding and extending in the usual manner. This woman's constant complaint was that during coition the penis seemed lost in the vagina and provided little direct exteroceptive stimulation during thrusting episodes.

There were two instances of an exceptionally small vagina within the study-subject group. These two women, when responding to plateau-phase levels of sexual tension, could accommodate either a large or a small penis, but experienced constant difficulty if mounting was attempted before sexual tensions had reached high levels of response. Early attempts frequently incited sufficient pain to dispel incipient or real sexual tensions. These women had difficulty accommodating any penis, regardless of its size, unless they were highly excited.

In the same category of potential distress are women who have undergone long periods of continence or are postmenopausal. In both situations there may be some shrinking of the vaginal barrel. These vaginas usually will respond slowly to stimulation at a first coital opportunity, reacting in fashion similar to that of the exceptionally small vagina described above. It becomes obvious that penile size usually is a minor factor in sexual stimulation of the female partner. The normal or large vagina accommodates a penis of any size without difficulty. If the vagina is exceptionally small, or if a long period of continence or involution due to aging intervenes, a penis of any size can distress rather than stimulate, if mounting is attempted before advanced stages of female sexual tension have been experienced.

Observation of the accommodation reactions within the human vagina has been made possible by the techniques of artificial coition developed during the past five years.

THE FUNCTIONAL ROLE

As a center of sensual focus, the clitoris serves both as a receptor and transformer of sexual stimuli (see Part 2 of Chapter 5). The penis is primarily a receptor of sexual stimuli, but there is also a transformer role for the penis in sexual response. The functional role of the penis is established through both exteroceptive and proprioceptive stimulation. As a receptor of sexual stimuli, the penis well may be as responsive to superficial stimuli as is the clitoris in the female. However, since the penis plays both active and passive roles in human sexual activity, as opposed to the more constant receptor or passive role of the clitoris, it is difficult to compare the two organs purely from a subjective standpoint.

Obviously the penis not only serves the male as a receptor and transformer of stimuli but also provides exteroceptive stimulation in its erect state for the female partner during coition. When the penis responds to sexual stimuli with an erective reaction, the physiologic fact of full erection and the proprioceptive stimuli that full erection develops for the male increases the pelvic focus of the male's sexual interest. The vasoconcentration that results in full penile erection serves in effect as a transformer mechanism by which the effectiveness of initial sexual stimuli is increased many times in the male's consciousness. As opposed to the female, in whom the clitoris in its psychophysiologic role of transformer projects the warmth of vasocongestive sensual demand throughout the pelvic viscera, the male's sensual focus usually is directed solely to the full, tense demand of the erect penile organ. Thus the penis serves as a psychophysiologic transformer of sensual stimuli through the proprioceptively stimulating process of erection. This is the first phase of the role of the penis as a transformer for sexual stimuli.

The second phase in the transformer role is developed by either vaginal or manual containment of the fully erect penile shaft. Regardless of whether the male is stimulated by intromission or by automanipulative techniques, the shaft of the penis is enclosed and thereby subjected to further sensory stimulation that primarily is exteroceptive in nature. The male's sexual tensions elevate from late excitement into plateau phase and toward ejaculatory demand as a result of proprioceptive stimuli plus successive contact and pressure stimuli created either by the thrusting of the engorged penile shaft within the accommodating vaginal barrel or by manual constriction of the penile shaft with manipulative techniques.

Both penile containment and thrusting provide stimuli, fundamentally exteroceptive in character, that are transformed through higher cortical centers into those levels of sensual focus that ultimately develop ejaculatory demand. The actual mechanism that triggers the male into an ejaculatory process is no better understood than that mechanism that provides the female with orgasmic release of her vasocongestion and myotonia. In both cases, however, the clitoris and the penis not only act initially as receptors for both objective and subjective sensual stimuli but also express their roles as transformers of sexual stimuli fundamentally by means of psychophysiologic orientation and responsiveness to proprioceptive stimuli.

AUTOMANIPULATION

During the discussion of the role of the clitoris in female sexuality, a clinical consideration of masturbatory techniques was presented (see Part 2 of Chapter 5). Similarly, a discussion of the penis in its functional role in male sexuality should include consideration of automanipulative techniques. Genital manipulation occurs in earliest infancy in both sexes. The pleasing sensation of genital play in infancy is translated by most men into the active pleasure of tension release engendered by genital manipulation during puberty or the teenage years. The instance of a positive masturbatory history was placed at 92 percent of the total male population by Kinsey and his associates [142]. Their figures generally have been supported in this country and abroad by many similar reports [57, 99, 117, 255, 269]. The age most frequently recalled by the 312 members of the male study-subject population for onset of active masturbatory practices centered around the fourteenth or fifteenth year. Some men described masturbatory patterns starting at the age of 9 or 10, others not until 16 or 18 years. However, by far the greatest onset frequency was concentrated during the immediate postpubertal years. It may be recalled that all members of both study-subject populations described a positive history of masturbatory facility (see Chapter 2).

Just as the female, males develop completely individual masturbatory techniques and overt response patterns. This despite the fact that a much higher percentage of boys observe their friends in masturbatory activity than do girls of similar age groups. Some men use the lightest touch on the ventral surface of the penis, some use strong gripping and stroking techniques that for many individuals would be quite objectionable, if not painful. Frequently men prefer stimulation of the glans alone, either confining manipulation to the ventral surface of the penis on or near the frenulum or using the simple finger technique of pulling at or stimulating the entire glans area. These are the exceptions, however, since most men manipulate the shaft of the penis with stroking techniques that encompass the entire organ and vary from man to man in desired rapidity, excursion, and tightness of manual constriction.

Uncircumcised males have not been observed to concentrate specifically on the glans area of the penis. Normally they follow the usual pattern of confining manipulative activity entirely to the penile shaft. Stroking techniques rarely move sufficiently distal on the shaft of the penis to encounter more than the coronal ridge of the glans even late in plateau phase just before ejaculation. For this reason the foreskin (even in those males with marked mobility of this tissue over the subjacent glans) rarely is retracted from the total glans area. Usually only that area of the glans immediately surrounding the urethral meatus is exposed prior to ejaculation. This is obviously a different picture from that occasioned by active intercourse. With full vaginal containment the foreskin not tightly attached to subjacent tissue usually retracts freely from a major portion of the glans during active male coital thrusting before ejaculation.

As the male reaches late plateau levels of sexual tension, the rapidity of manipulative excursion increases, until most men are stroking the penile shaft as rapidly as possible. However, with the onset of the ejaculatory process marked variation has been observed in male manipulative technique. During ejaculation most of the study subjects either cease completely or markedly slow the manual excursion along the penile shaft. Many of the study subjects grip the penile shaft spastically (usually just beneath the glans) and continue this spastic constrictive pressure during the entire ejaculatory process. This reaction may represent an involuntary response pattern similar to that developed during active coition. Many men plunge the penis into deepest possible vaginal containment as ejaculation develops and cease all active pelvic thrusting during the ejaculatory process.

A few study subjects manipulate the penile shaft actively during the ejaculatory process, although almost universally slowing the rapidity of and easing the constrictive tension of the stroking techniques. These men observed during active coition usually avoid spasmodic, deep vaginal containment of the penis with onset of ejaculation and continue coital thrusting during their entire ejaculatory process. No accurate check has been made of this reactive mannerism, but it is estimated that not more than 10 percent of the male study-subject population continued active stroking, either manipulative or coital in origin, during their ejaculatory response.

Many men have reported the penile glans to be quite sensitive to any pressure or containment immediately subsequent to ejaculation. A few of the study subjects develop such a degree of glans sensitivity that they involuntarily protect it against any form of stimulation. These men reject any pattern of continued intravaginal containment after ejaculation because even the low-grade exteroceptive stimulation of the static vaginal barrel is sufficient to distress. Certainly, care is taken by males with postejaculatory glans sensitivity to avoid any continuation of penile stroking after automanipulative orgasm. The sensitivity of the penile glans (like the sensitivity of the clitoral glans) in the immediate postorgasmic period rarely is appreciated by the opposite sex. Significant clinical distress develops for these individuals when the male attempts continued stimulation of his female partner or the female continues active manual stroking or pelvic thrusting immediately subsequent to the male's ejaculation. Vocalization of postorgasmic glans sensitivity in either marital partner will clear up this potential clinical distress in short order.

From the point of view of sensate focus, the male usually is not as effectively stimulated sexually by manipulation of the scrotal sac and its subjacent content as is the female from stimulation of the labial area or the vaginal introitus. While male study subjects responded during the neurologic examination to the lightest touch of the scrotal sac to the same degree that they did to similar stimulation of the penile shaft, they have not described comparable sensual pleasure derived from stimulation of the two areas. Obviously, this lack of scrotal sensual focus is at variance with the focus of sensuality returned from stimulation of the labial areas and the vaginal introitus expressed by the women examined by the Kinsey gynecologists [144].

It is obvious that the sensual focus of the entire male reproductive viscera is limited primarily to the penile shaft and glans. This is opposed to the female, who has not only a high level of sensual focus concentrated in the clitoral shaft and glans but also major levels of sensual focus in the labia, at the vaginal outlet, and in the vaginal barrel. The sensitivity of the rectum to stimulation was adjudged essentially equal between the two sexes by gross clinical observation. It must be remembered, however, that material of homosexual content has not been included in this review.

FEARS OF PERFORMANCE

"Phallic fallacies" relating to the functional role of the penis frequently devolve from the culturally conceived role for the male partner in human coition—that of actively satisfying the female partner.

The "fear of performance" developing from cultural demand for partner satisfaction has been in the past uniquely the burden of the responding male. Inevitably fear provides a breeding-ground for misconception. Among the male members of the study-subject population and males interrogated as applicants, phallic fallacies of subjective orientation were related to decades of life experience more than to any other single factor.

Since a criterion for membership in the study-subject population was ejaculatory experience in both masturbatory and coital situations, no members of the study group had to contend personally with performance fears developed from primary impotence. Questioning related to this type of subjective concern among this selective population exposed beliefs that primary impotence was only the result of a complete homosexual orientation or elicited vague replies that "something must be wrong with the glands." Delusions resulting from the fears of performance with which the primarily impotent male must contend almost on a daily basis were undeveloped to the total group of interrogated males and, therefore, cannot be dealt with objectively.

For the men forty years or younger, fears of performance centered about questions of excessive ejaculatory experience and concerns for premature ejaculation. The problem of too frequent ejaculation was associated in the minds of many study subjects with possible loss of physical strength and not infrequently was presumed to be a basis for emotional instability if not severe neurosis. These misconceptions have grown from the culturally centered fear that frequent or excessive masturbation may lead to mental illness. No study subject could provide a secure personal concept of what constituted frequent or excessive levels of masturbation, nor could anyone describe an instance known to them, even by report, of mental illness resulting from masturbation. The superstition that physical or mental deterioration results from excessive masturbation is firmly entrenched in our culture, if returns from the team questioning of the total male group of study-subject applicants are any criterion.

Reported masturbatory frequency in the male study-subject group ranged from once a month to two or three times a day. Every male questioned expressed a theoretical concern for the supposed mental effects of excessive masturbation, and in every case "excessive levels" of masturbation, although not defined specifically, were considered to consist of a higher frequency than did the reported personal pattern. One man with a once-a-month masturbatory history felt once or twice a week to be excessive, with mental illness quite possible as a complication of such a frequency maintained for a year or more. The study subject with the masturbatory history of two or three times a day wondered whether five or six times a day wasn't excessive and might lead to a "case of nerves." No study subject among the 312 questioned in depth expressed the slightest fear that his particular masturbatory pattern was excessive regardless of stated frequency.

There is no established medical evidence that masturbation, regardless of frequency, leads to mental illness. Certainly there is no accepted medical standard defining excessive masturbation. It is true, of course, that many severely neurotic or acutely psychotic men masturbate frequently. If a high-frequency pattern of masturbatory activity exists, it may be but one of a number of symptoms of underlying mental illness rather than in any sense the cause of the individual distress. The vague concept of excessive masturbatory activity is a phallic fallacy widely accepted in our culture, relating specifically to the functional role of the penis in male sexuality.

Problems of premature ejaculation also disturbed the younger members of the study-subject population. These fears of performance were not associated with problems of erection; rather, they were directed toward the culturally imposed fear of inability to control the ejaculatory process to a degree sufficient to satisfy the female partner. These expressed fears of performance were confined primarily to those study subjects who had attained college or postgraduate levels of formal education. Only 7 of the total of 51 men whose formal education did not include college matriculation expressed the slightest concern with responsibility for coital-partner satisfaction. These men felt that it was the female's privilege to achieve satisfaction during active coition if she could, but certainly it was not the responsibility and really not the concern of the male partner to concentrate on satisfying the woman's sexual demands. Out of a total of 261 study subjects with college matriculation, 214 men expressed concern with coital-partner satisfaction. With these men ejaculatory control sufficient to accomplish partner satisfaction was considered a coital technique that must be acquired before the personal security of coital effectiveness could be established.

The fear of performance reflecting cultural stigmas directed toward erective inadequacy was that associated with problems of secondary impotence. These fears were expressed, under interrogation, by every male study subject beyond forty years of age, irrespective of reported levels of formal education.

Regardless of whether the individual male study subject had ever experienced an instance of erective difficulty, the probability that secondary impotence was associated directly with the aging process was vocalized constantly. The fallacy that secondary impotence is to be expected as the male ages is probably more firmly entrenched in our culture than any other misapprehension. While it is true that the aging process, with associated physical involution, can reduce penile erective adequacy, it is also true that secondary impotence is in no sense the inevitable result of the aging process. This factor has been firmly established by the aging members of the study-subject population (see Part 2 of Chapter 16). In addition, the clinical-research program of the Reproductive Biology Research Foundation has established the fact that most secondary impotence associated with the aging process can be transitory in character.

In most instances, secondary impotence is a reversible process for all men regardless of age, unless there is a background of specific surgery or physical trauma.

The functional role of the penis in male sexuality has not been established with the security of the organ's functioning role. It has been severely obscured by "phallic fallacies" of cultural origin. Further definitive research in the physiology of male sexual response will make the greatest contribution toward identifying and reversing these misconceptions. It is inevitably true that the psychology of human sexual response can best be appreciated when the physiology has been established. The numerous "phallic fallacies" dealt with in this chapter are but a few of the many present in our culture. They represent major concerns of a highly selective group of male study subjects and applicants, and not necessarily the fears of a cross-section of the male population.

13

THE SCROTUM AND THE TESTES

ANATOMY AND PHYSIOLOGY

The scrotum and the testes respond to sexual stimulation, as do all other male primary and secondary organs of reproduction, with both localized vasocongestion and increased myotonia. In a sexually unstimulated state the thin scrotal integument (skin and tunica dartos) characteristically hangs in multiple folds or creases and moves freely over the subjacent scrotal content. The scrotum is morphologically homologous to the labia majora of the human female. However, there is not the mature female's tendency toward localized deposition of fatty tissue.

THE SCROTUM

The scrotum responds to sexual stimulation in specific reaction patterns that reflect the severity of the stimulative response. Therefore, descriptions of these reaction patterns can be related to the four phases of the cycle of sexual response.

EXCITEMENT PHASE

As sexual tensions rise there are notable tensing and thickening of the scrotal integument accomplished both by localized vasocongestion and by contraction of the smooth-muscle fibers of the dartos layer (see Fig. 12-4). The unstimulated scrotal patterns of multiple folding and free movement rapidly are lost. Constriction of the scrotal integument produces a significant decrease in the internal diameter of the scrotal sac. There also is a marked restriction in free testicular movement in other than a perpendicular plane.

The constricted scrotal sac with its resultant loss of internal diameter contributes secondary support to the reaction of testicular elevation during advanced stages of the excitement phase. If excitement-phase levels of sexual tension are maintained for long periods without the usual progression toward plateau-phase levels of intensity, the constricted and congested scrotal sac may relax. When relaxation occurs scrotal folding reappears, and partially elevated testes may return to the scrotal depths. This loss of scrotal physiologic responsiveness may develop even though full penile erection is maintained. Ultimately, if sexual tensions accumulate and plateau-phase levels of response develop, the congestive and constrictive reactions of the scrotal sac may be reestablished rapidly.

PLATEAU AND ORGASMIC PHASES

Occasionally an obvious thickening and tensing of the scrotal integument develops as a continuing reaction into plateau-phase levels of sexual tension, but only if the excitement phase has been of exceptionally short duration (see Figs. 12-5, 12-6). There are no specific scrotal sac reactions to either plateau- or orgasmic-phase levels of sexual tension.

RESOLUTION PHASE

During the resolution phase the scrotal integument responds to loss of sexual tensions in two diametrically opposed patterns. Either reaction pattern usually is specific for the individual male. The most frequent reaction is characterized by rapid loss of the congested, tense appearance of the scrotum and an early reappearance of the patterns of integumental folding or creasing so characteristic of the sexually unstimulated male (see Fig. 12-7). Approximately 25 percent of human males demonstrate a pattern of slow resolution with delayed loss of the scrotal integument's tension-induced congestion and constriction. Retained definitive thickening of the integument frequently has been observed to last for one to two hours after an ejaculatory experience. The concomitant constriction of internal diameter of the sac may delay (five to twenty minutes) bilateral testicular return to fully relaxed positioning in the scrotal depths.

As stated, the individual male usually adheres to either rapid or slow scrotal involution, but three individuals have demonstrated both response patterns. For these three men, the longer that tension-induced congestion and constriction of the scrotal sac are maintained before orgasmic release, the more slowly detumescence and relaxation of the scrotal sac will develop during the resolution phase of the sexual response cycle.

No specific color change of the scrotal integument has been observed during any phase of the sexual response cycle.

THE TESTES

EXCITEMENT PHASE

The testes evidence specific reaction patterns during each of the four phases of the sexual response cycle. As excitement-phase levels of tension develop, there is a specific elevation of both testes toward the perineum (see Fig. 12-4). This physiologic response to sex tension increment has been observed during previous investigations [3, 144, 213]. Testicular elevation is accomplished by shortening of the spermatic cords. The cremasteric musculature which invests the spermatic cord produces this physiologic response through an entirely involuntary contractile process.

Actually, only partial elevation of the testes is accomplished during the excitement phase, unless there is to be a fulminating completion of the sexual response cycle. The testes in addition to their ascent toward the perineum undergo a change in their axis of suspension during the process of elevation. The superior pole of the ascending testis rotates anteriorly, and as a consequence, once the testis is elevated completely, the posterior testicular wall ultimately comes in direct contact with the male perineum. Actually this modest (30° to 35°) anterior rotation has onset late in the excitement phase and is completed during the terminal portions of the plateau phase of the individual sexual response cycle.

If excitement-phase levels of sexual tension are maintained for more than 5–10 minutes and there is no immediate interest in escalating the sexual tensions to plateau-phase levels, the contracted cremasteric musculature of the spermatic cords relaxes and the testes together with the relaxed scrotal sac (see previous description) return to their unstimulated suspended positioning. Cremasteric muscular tension usually cannot be maintained over indefinite periods of time. Therefore, the testes frequently may return to their unstimulated low-scrotal positioning despite the fact that excitement-phase levels of sexual tension are maintained, and an accompanying penile erection may be continued indefinitely. The sexually responding male may go through several episodes of partial elevation and subsequent descent of the testes during a voluntarily extended excitement phase before sufficient sexual tensions accumulate to establish a plateau-phase level of sexual response and with it the resultant complete testicular elevation.

Observation of the phenomenon of preejaculatory testicular elevation finally has provided an adequate explanation of the physiologic intent of the well-established clinical entity of the cremasteric reflex [162]. However, it should be emphasized that partial testicular elevation should not be considered purely as a physiologic response to sexual tension. Under controlled laboratory conditions, testicular elevation (approximately half-way to the perineum) has been observed as a response to a chilled environment and also has been produced as an immediate response to the stimuli of both fear and anger.

PLATEAU PHASE

As male sexual tensions rise through plateau-phase toward orgasmic-phase release, the specific reaction of testicular elevation progresses until the final preejaculatory positioning in tight apposition to the male perineum is attained (see Fig. 12-5). Repeated demonstrations have established the fact that the phenomenon of testicular elevation is of extreme physiologic importance. If the testes do not undergo at least partial elevation the human male will not experience a full ejaculatory sequence. Some males, particularly after the age of fifty, do ejaculate from a partially rather than a fully elevated testicular positioning. When ejaculation from partially elevated testicular positioning occurs, there is marked reduction in ejaculatory pressure. This variation in response pattern will be discussed in detail during the consideration of the influence of age upon the human male's sexual response patterns (see Part 1 of Chapter 16).

When the testes do rise to a position of close apposition to the male perineum, an orgasmic phase is certain to follow if effective sexual stimulation is maintained. Full testicular elevation is pathognomonic of impending ejaculation.

The left testicle, which usually (85 percent of the time) has the lower scrotal sac positioning in a sexually unstimulated state, obviously must move through the widest excursion during testicular elevation. The left testis also has been observed to react frequently to the stimulus of sexual tensions independently of the right testicle. Many males accomplish final elevation of the right testicle against the perineum late in excitement or early in a plateau-phase sequence, while the left testicle still continues to move up and down in restricted excursion from partial to complete elevation within the congested and constricted scrotal sac. The left (lower) testicle frequently does not establish full perineal apposition until immediately prior to the ejaculatory sequence.

There is yet another physiologic response of the testicles to elevated sexual tensions that has not been described heretofore. This reaction consists of an observable increase in testicular size. Direct palpation of the testes supports the clinical impression that this is yet another deep vasocongestive phenomenon. Testicular size increase usually is not apparent until late excitement or early plateau phase. The testes achieve approximately a 50 percent size increase over their sexually unstimulated noncongested state, before the orgasmic phase is experienced. Some males have been observed to develop almost a 100 percent increase in testicular size, particularly if the individual sexual response cycle is of exceptional duration. As a rough rule of thumb, it can be stated that the longer the plateau-phase levels of sexual tension are maintained without orgasmic phase release, the more severe is the deep vasocongestion of the testes and the more obvious is the resultant testicular size increase.

ORGASMIC PHASE

No specific orgasmic phase reaction of the testes has been recorded (see Fig. 12-6). Suspicion persists, however, that this is merely an expression of lack of effective observation and physiologic recording techniques rather than a factual determination. Specifically, the testes have not been felt nor observed to contract during the ejaculatory process. To date, finite physiologic recording of testicular contractile potentials has not been attempted.

RESOLUTION PHASE

Resolution-phase loss of vasocongestive increase in testicular size and full descent of the testes into the depths of a relaxed scrotum (see Fig. 12-7) may develop as either a rapid or a slow involutionary pattern, just as has been described for the scrotal sac relaxation. There is marked specificity of response pattern for the individual male, but wide variation of involutionary reaction from man to man. However, the general clinical pattern of target-organ vasocongestion again applies. The longer the plateau-phase levels of sexual tensions are experienced and resultant testicular size increase is maintained, the slower is full testicular detumescence during the resolution phase.

The specific physiologic reactions of scrotum and testes to sex tension increment have been considered in this chapter. Again evidence has been developed of both superficial (scrotal integument) and deep (testicular size increase) vasocongestion and both generalized (scrotal constriction) and specific (testicular elevation) myotonia. Application of this material will be developed in a future publication relating to clinical consideration of problems of human sexual inadequacy.

I 4

THE MALE ORGASM
(EJACULATION)

The human male's orgasmic experience (ejaculation) can be approached from the same three disciplinary foci that have been employed in Chapter 9 in an attempt to interpret the human female's orgasmic experience. These foci are: (1) physiologic (characteristic physical conditions and reactions during the peak of sex tension increment); (2) psychologic (psychosexual orientation and receptivity to orgasmic attainment); and (3) sociologic (cultural, environmental, and social factors influencing orgasmic incidence or ability) [12, 145, 322].

In the past the progressive chain of physiologic events that comprise the human male's orgasmic episode has been considered in detail. However, there has been little attempt to present an analysis of psychologic components and to correlate objective and subjective elements of the experience. During the cycle of human sexual response, the male reaction for which there is no comparable female counterpart is, of course, the emission of seminal fluid. The physiology of this orgasmic process has been established and will be reviewed but briefly. The male's subjective progression through orgasm will be considered in more depth as correlation between physiologic and psychologic components of the experience is attempted. Sociologic aspects of male orgasm will be examined in context.

Preliminary to the discussion of male orgasmic response, a characteristic plateau-phase reaction should be mentioned. Frequently a preorgasmic secretory emission has been observed. It is mucoid in character, usually totals no more than two or three drops, and escapes involuntarily from the urethral meatus. Occasionally, developing as an individual reactive characteristic or during long-maintained plateau-phase tension levels, the mucoid emission has reached levels of 0.5–1 cc. in total volume. Frequently, actively motile spermatozoa have been demonstrated in microscopic examinations of this preejaculatory fluid emission.

There has been no attempt to establish percentage incidence for the preejaculatory material in the male study-subject population because many males secrete the material so irregularly. As stated, it appears most frequently during voluntarily lengthened plateau-phase experiences. For example, in active coition a man may practice voluntary ejaculatory control at plateau-tension levels through several of his female partner's orgasmic cycles. Other similar situations tend to increase both frequency of occurrence and secretory volume of the preejaculatory mucoid material. They are automanipulative activity voluntarily maintained at plateau-phase tension levels for lengthy periods without ejaculatory release, and fellatio conducted in similar manner and with similar intent.

From a physiologic point of view, it is interesting to note that the plateau-phase timing of the male's preejaculatory emission is essentially the same as the plateau-phase secretory activity of Bartholin's glands in the female sexual cycle. The source of the mucoid material produced by the sexually responding human male has been allocated questionably to Cowper's glands (see Fig. 12-3). The chemical composition of either of these plateau-phase secretions has not been identified.

No relationship has been established between incidence or amount of the preejaculatory mucoid material and the volume or specific content of the true seminal emission. To date, there is insufficient information to correlate the incidence of large numbers of active spermatozoa in the preejaculatory secretion with high levels of fertility as reflected by in vitro evaluations of seminal-fluid content. There also is no statistically secure parallel between the presence or absence of a preejaculatory emission and a higher or lower seminal plasma volume [131]. In short, the physiology and biochemistry of the male's preejaculatory emission are essentially unexplored areas.

PHYSIOLOGY OF ORGASM

The actual expulsion of seminal-fluid content from the accessory or secondary organs of reproduction (prostate, seminal vesicles, ejaculatory duct, etc.) into the prostatic urethra and the progression of the fluid content under pressure through the full length of the penile urethra to the urethral meatus are the physiologic expression of male orgasmic experience. The act of ejaculation usually involves activity of the highest cortical levels but also may be purely reflex [110, 168, 206, 260, 285]. The ejaculatory process may be divided physiologically into two separate stages. Stage I consists of expulsion of seminal-fluid substrate from the accessory organs of reproduction into the prostatic urethra (see Fig. 12-3). Stage II includes the progression of seminal-fluid content from the prostatic portion of the urethra through both the membranous and the penile segments of the urethra to the urethral meatus (see Fig. 12-1).

STAGE I

The first stage of the ejaculatory process is initiated by accessory-organ contractions previously described as commencing with the vasa efferentia of the testes [224]. These secondary contractions are presumed to continue through the epididymis to the vas deferens which finally contracts in close parallel with the seminal vesicles [225]. Regularly recurring prostatic contractions have been palpated rectally in study subjects during ejaculation. This physiologic reaction of the prostate gland also has been recorded by an independent observer [265].

As many as six separate fractions of the seminal fluid have been identified [284]. Initially, seminal-fluid content is developed from expulsion of prostatic fluid into the prostatic urethra. Then contents of the ampulla (outlet of the vas deferens) are discharged into the prostatic urethra, accompanied by simultaneous expulsion of seminal-vesicle content. During the course of the ejaculatory episode prostatic fluid is delivered to the seminal-fluid content repeatedly by regularly recurring contractions. As stated, these expulsive contractions of the prostate can be palpated rectally.

As seminal fluid collects in the prostatic urethra, there simultaneously is a two- to threefold involuntary expansion of the urethral bulb (see Fig. 12-8). This expansion develops in anticipation of the second-stage expulsive urethral contractions.

At the onset of the ejaculatory experience, the intact internal sphincter of the urinary bladder normally closes or remains sealed, thereby preventing seminal-fluid content from entering the bladder in a retrograde ejaculatory sequence. This specific internal-sphincter constriction also has the function of retaining any urinary content within the bladder and avoiding any intermingling of urine and seminal plasma [120].

STAGE II

The second stage of the ejaculatory process is initiated by relaxation of the external sphincter of the bladder, which allows the seminal-fluid content to flow into the distended bulb and penile urethra. The seminal plasma is propelled from the prostatic urethra along the penile urethra by the perineal musculature, the bulbospongiosus and ischiocavernosus muscles, and the sphincter urethrae. The urethral bulb also contracts regularly as an aid to the propulsive mechanism. The first two or three ejaculatory contractions of the penile urethra project seminal-fluid content under such pressure that initial portions of the ejaculate may be expelled 12–24 inches from the urethral meatus if the penis is unencumbered by vaginal containment. The anatomy of seminal-fluid deposition within the vaginal barrel is discussed in Part 2 of Chapter 6.

Alterations in ejaculatory physiology reflecting the influence of the aging process are presented in Part 1 of Chapter 16.

The physiology of repeated ejaculation should be considered briefly. Many males below the age of 30, but relatively few thereafter, have the ability to ejaculate frequently and are subject to only very short refractory periods during the resolution phase. One male study subject has been observed to ejaculate three times within 10 minutes from the onset of stimulative activity. The seminal-fluid volume progressively was reduced in amount with each ejaculatory episode. This example, of course, marks the exception to the basic rule of severe male psychophysiologic resistance to sexual stimuli (refractory period) immediately after an ejaculatory experience (see Fig. 1-1).

From the standpoint of physiologic strain, there is no information available to date that would indicate any residual physical distress from repeated ejaculation that might be expected to develop in a healthy male. The act of ejaculation, while obviously a part of the total of physiologic strain of orgasmic experience, has not been observed to create residual physical distress. There is a widespread concept that ejaculation, whether accomplished through masturbation or coition, is detrimental to the physical condition of men in athletic training programs. To date, there is no physiologic evidence to support this concept.

From a physiologic point of view the male orgasmic experience is one of total-body involvement through the processes of vasocongestion and myotonia. Specific orgasmic reactions of body areas and organ systems have been outlined in Chapter 11 and need not be repeated here. It is important, however, to reemphasize constantly the protean character of orgasmic experience in the male as well as the female. The male's ejaculatory reaction frequently draws attention from, and tends to obscure the degree of, total-body involvement developed by an orgasmic interlude.

PSYCHOLOGY OF ORGASM

The two-stage physiologic process of ejaculation can be correlated specifically with the male's subjective progression through the orgasmic experience.

Subjective material has been returned from interrogation of 417 males with ages ranging from 18 to 89 years. Many of these men (164) have been or are actively with the research program as cooperating study subjects. Eighty-nine other men have been study-subject applicants. Additionally, material has been drawn from 65 sexually inadequate males seeking relief from their clinical distress. Finally, 99 infertile males were selected from the conceptive-physiology research program. Their qualifying factor was physiologic evidence of excessively high or low ejaculate volume. The discussion to follow represents a consensus of the opinions expressed and sensations described by these voluntarily cooperative males.

STAGE I

In the human male a sensation of ejaculatory inevitability develops for an instant immediately prior to, and then parallels in timing sequence, the first stage of the ejaculatory process (accessory-organ contractions). This subjective experience has been described by many males as the sensation of "feeling the ejaculation coming." From onset of this specific sensation, there is a brief interval (2 to 3 seconds) during which the male feels the ejaculation coming and no longer can constrain, delay, or in any way control the process. This subjective experience of inevitability develops as seminal plasma is collecting in the prostatic urethra but before the actual emission of seminal fluid begins. The two- to threefold distention of the urethral bulb developing in the terminal portions of the plateau phase also may contribute proprioceptively to the sensation of ejaculatory inevitability [213, 290].

STAGE II

During the second stage of the ejaculatory process (propulsion of seminal-fluid content from prostatic urethra to the urethral meatus), the male subjectively progresses through two phases: First, a contractile sensation is stimulated by regularly recurring contractions of the sphincter urethrae. Second, a specific appreciation of fluid volume develops as the seminal plasma is expelled under pressure along the lengthened and distended penile urethra.

Phase 1. The contractile effect varies in intensity of subjective appreciation from onset to termination of the expulsive process. The severity of the first two or three expulsive contractions of the penile urethra and the slowed, almost tensionless, final contractions of the ejaculatory process create entirely different contractile sensations. The first few forceful contractions frequently develop a relative degree of secondary anesthesia along the barrel of the penile urethra, so that the final portions of the seminal volume propelled by relatively tensionless contractions may escape without the male's sensate awareness of the emission. When the male is subjectively aware of the final tensionless contractions, there is no associated level of pleasure response similar to that identified with the first strong expulsive contractions.

Phase 2. The subjective appreciation of volume of seminal-fluid content is best exemplified by the severe orgasmic experience related to an initial ejaculation after a period of continence as opposed to the subjectively milder orgasmic episode associated with a second ejaculation developing in a short interval after a first emission. If a male has been continent for several days, there generally is a larger volume of seminal fluid ejaculated compared to that returned after a few minutes of continence [131, 187]. The larger fluid volume is appreciated subjectively as a more sensually pleasurable sensation than is the lower volume ejaculate. Obviously, there must be recurrent ejaculations for the male subjectively to appreciate volume differences.

There is more to learn of the male's subjective appreciation of seminal-fluid volume. A larger ejaculate volume may account in part for the male's relatively greater pleasure in an initial ejaculatory episode after a significant period of continence than in a repeated orgasmic experience at the termination of his first refractory period. This subjective reaction pattern is in opposition to reported orgasmic response patterns for the human female. When female study subjects were interrogated in the laboratory after multiorgasmic experiences, the second or third orgasmic episode usually was identified subjectively as more satisfying or more sensually pleasurable than the first orgasmic episode. When male study subjects were multiejaculatory in the laboratory, inevitably the first ejaculatory episode was reported as the most satisfying experience.

It should be emphasized that the first-phase sensation of contractile response and the second-phase appreciation of fluid volume blend as the second stage of the male's ejaculatory experience progresses. The contractile sensation is the dominant of the two factors, as it is experienced initially and continues to be of subjective import during and after seminal-fluid volume appreciation has developed and subsided. The final sensate focus in the ejaculatory experience is on contractions of the penile urethra recurring irregularly and with rapidly diminishing intensity.

The subjective progression of the two phases of contraction and fluid volume through Stage II of the male orgasmic experience is directly comparable to the sensations of contraction and throbbing that form the two phases of Stage III of the human female's subjective progression through orgasmic experience (see Chapter 9).

There are marked differences in both objective and subjective orgasmic experience for the aging male as opposed to his younger counterpart. These differences have been considered in detail in Chapter 16.

In contrast to the fact that orgasmic experience of the human female can be interrupted by extraneous psychosensory stimuli, the male orgasmic experience, once initiated by contractions of the accessory organs of reproduction, cannot be constrained or delayed until the seminal-fluid emission has been completed. Regardless of intensity of extraneous sensory stimuli, the male will carry the two-stage ejaculatory process to completion.

SOCIOLOGIC FACTORS IN ORGASMIC ACHIEVEMENT

As opposed to the evasive literature reflecting sociologic influences upon the female orgasmic expression, there is little literary concern for the male's orgasmic experience. There are two major reasons for this lack of sociologic concern with ejaculation. Of primary importance is the fundamental demand of the life cycle for male ejaculation. This one factor has provided acceptance of the ejaculatory process per se by all cultures. Obviously, there have been cultural attempts to control ejaculatory frequency and to direct ejaculatory occasion, but not to repress the ejaculatory process. This one factor of ejaculatory necessity has relieved the male of the psychosocial pressures that have been imposed upon the female's orgasmic experience.

The second reason for lack of sociologic concern with the male's orgasmic experience is the fact that cultural pressures have been directed toward other target areas. For the male, these pressures have centered about the physiologic processes of penile erection and not ejaculation. Thus, cultural demand has played a strange trick on the two sexes. Fears of performance in the female have been directed toward orgasmic attainment, while in the male the fears of performance have related toward the attainment and maintenance of penile erection, and orgasmic facility always has been presumed.

It is evident that man's sexual inadequacy is not related directly to his ability or inability to attain orgasmic release of sexual tensions. Psychosocial influences certainly create clinical states of male sexual inadequacy, but rarely are they directed specifically toward the orgasmic experience. Rather the major clinical distresses of primary and secondary impotence obviously have direct relation to the psychophysiologic concerns of attainment and/or maintenance of penile erection and do not relate to the actual ejaculatory process. It also is obvious that the psychosocial concern of premature ejaculation, although directly related to the male's orgasmic episode, hardly represents an expression of orgasmic inadequacy. Therefore, these physiologic reflections of psychosocial imbalance, arbitrarily termed sexual inadequacies, have been considered in Part 2 of Chapter 12 in a brief discussion of the clinical concerns of the penile erective process.

There is a rare clinical exception to the concept that male sexual inadequacy culturally is not related directly to orgasmic attainment. Five men have been referred to the conceptive-physiology section of the reproductive-biology research program during the last 18 years with the primary complaint of conceptive inadequacy. These five men have not had difficulty in erective attainment, nor has there been any inadequacy in maintenance, once penile erection was achieved. In fact, their difficulties are exactly on the opposite side of the coin. Their conceptive and psychosocial problems have centered on the physiologic fact that they cannot ejaculate with the penis contained in the vagina. These men can and do maintain coital connection for 30 to 60 minutes at any given opportunity, but they are not able to ejaculate intravaginally.

Three of the five wives are multiorgasmic as result of the constant opportunity at long-maintained coition. Coital connection is terminated by the female partner's admission of sexual satiation. The remaining two wives, although having proved responsive capacity, had some loss of responsive interest developing from concern about the considered partner abnormality.

All five of these men, ranging in age from 28 to 41 years, have a masturbatory history reflecting some regularity of automanipulative or partner release. In addition, they report occasional nocturnal emissions. Of interest is the fact that they describe only excitement-phase levels of sexual tension developing during or immediately after the extended coital episodes. They find little or no interest in tension release associated directly with coition. Usually their psychosexual tension demands elevate and are expressed as completely separate sexual episodes. Their ejaculatory demand rarely is above three or four times a month. Only one of the five men has described four occasions of ejaculatory success with women other than his marital partner. There has been no more than one episode of success with each woman despite repeated attempts. Three of the men have had no ejaculatory success with other partners, although multiple exposure is described, and one man has denied extramarital experimentation. Positive homosexual histories have been obtained from only two of the five men, with only one of the men active at the time of consultation with the research program.

These men and their wives were referred initially because of problems of conceptive inadequacy. The primary marital-unit concern was for conception, not ejaculation. Since these problems have been resolved in three of the five families by using the husband's seminal content in insemination techniques, these men remain of interest more from a psychosexual point of view than as problems in conceptive physiology.

While the psychosocial implications of this relatively rare instance of male orgasmic inadequacy are striking, they have no place in the current restricted discussion. These clinical problems, together with those of primary and secondary impotence and premature ejaculation, will be discussed in detail in future publications directed toward psychosocial background and the diagnosis and treatment of human sexual inadequacy. Suffice it to say that these five men prove exceptions to the basic concept that the human male's failures in sexual expression rarely have psychologic or physiologic focus on the actual orgasmic (ejaculatory) experience. Inherent in the expression of this concept is the major difference in the psychosocial approach of our culture to male and female sexual inadequacy.

In essence, orgasm for the male, a two-stage experience, can be identified by a chain of specific physiologic reactions and by correlated patterns of subjective progression. Cultural concerns for male sexual performance do not focus on orgasmic attainment.

GERIATRIC SEXUAL RESPONSE

I5

THE AGING FEMALE

1 . ANATOMY AND PHYSIOLOGY

The anatomy and physiology of female sexual capacity and performance during and after the menopausal years have not been investigated previously. As might be expected, the cooperation of women in this age group is not elicited easily. It will require at least another decade to obtain the cooperation of aging women in numbers sufficient to provide biologic data of statistical significance. Current material is presented to suggest clinical impression rather than to establish biologic fact.

The number of older women (menopausal and postmenopausal) who have cooperated with the overall investigative program are listed and separated into ten-year age groups in Table 15-1. The oldest woman in the female population was 78 years at the time of her evaluation. Altogether, 61 women past 40 years (34 of these were past 50 years) of age have cooperated with the investigative program during the past decade.

Since patterns of sexual response have been established for premenopausal women using the four phases of the sexual cycle as an arbitrary descriptive mechanism, older women's sexual response patterns will be described in similar fashion. Physiologic variations from younger women's established reaction patterns will be emphasized in context.

This technique of comparing the sexual reactions of older and younger women should not be presumed to suggest physiologic abnormality for the reactions of older women. Norms of sexual response have been established independently for aging women without regard to reactive potentials of younger women. The technique of age-group comparison will serve merely to emphasize changes in the physiology of sexual response that are related to

TABLE 15-1

Age Distribution of 61 Female Menopausal
and Postmenopausal Study Subjects

Age Distribution	No. Active Participants	
41–50	27	
51–60	23	
61–70	8	34
71–80 *	3	
Total	61	

* Oldest study subject was 78 years old.

the aging process. The primary purpose of this report is to highlight the previously undescribed sexual response patterns of the aging human female.

EXTRAGENITAL REACTIONS

THE BREASTS

Excitement Phase. Nipple erection occurs in the aging female following exactly the patterns described for her younger counterpart (see Chapter 3). This reaction is the first external evidence of elevated sexual tensions and, presuming the nipples are not inverted, occurs shortly after the onset of any form of effective sexual stimulation. Members of the 60- and 70-year age groups demonstrated facility in the nipple erective response, just as their younger counterparts always have done. Apparently, this elastic-tissue activity is not destroyed by the aging process. There are, of course, smooth-muscle fibers in the nipple that contribute to the erective reaction, but the facility of nipple response cannot be credited to smooth-muscle contractility alone.

The vasocongestive increase in breast size, often evident under sex tension influence in the younger female who has not suckled, undergoes progressive involution in reactive effectiveness as the human female ages. Sixteen of the 27 members of the 41–50-year age group repeatedly demonstrated obvious increase in breast size as excitement-phase levels of sexual tension were established. Of this group of 16 women, only 4 had suckled babies. However, in the 51–60-year age group there was a marked reduction of the vaso-congestive reactive potential of the breasts. Only 5 of the 23 members of this age group demonstrated a clinically obvious increase in breast size during the excitement phases of their sexual response cycles. Yet 15 of the 23 members of this age group had not nursed. None of the 11 women over the age of 60 showed any clinically obvious increase in breast size as their sexual tensions mounted, and 6 of these 11 women gave no history of prior suckling. Thus, the clinical impression has been created that as the human female ages, some degree of the normal vasocongestive reaction of the breasts to elevated sexual tensions is delimited by the aging process. Had the study subjects been younger women, there would have been a much higher per-centage of vasocongestive increase in breast size, particularly among those women who had not suckled babies.

In those aging women that demonstrated obvious increment in breast size, the swollen superficial venous patterns and slow increase in breast volume (ultimately one-fifth to one-fourth size increase) followed exactly the reaction patterns of the younger age groups.

Plateau Phase.   Engorgement of the areolae, a constant finding during late excitement and early plateau phases of sexual response in younger women, also develops in their aging counterparts. How-ever, the intensity of the reaction usually is diminished. In younger women plateau-phase areolar tumescence is of such magnitude that it impinges upon the fully erect nipple, giving the impression that a significant degree of nipple erection is lost. Although areolar tumescence develops in women past 50 years of age, the reaction is of minor intensity and there is no impression of loss of nipple erection. Of clinical interest is the fact that women beyond 50 years of age may demonstrate an areolar tumescent reaction in one breast and not the other. This phenomenon has been observed rarely in younger women, but has been seen frequently in the older age group.

Fourteen of the 27 members of the 41–50-year age group showed a pink mottling over the anterior, lateral, and/or inferior surfaces of the breasts immediately prior to orgasmic release of their sexual tensions. Three of the 23 members of the 51–60-year age group reacted in similar fashion. None of the women past the age of 60 demonstrated the sex tension flush over the breasts during the plateau phase of the sexual cycles.

Orgasmic Phase. There is no specific breast reaction to the experience of orgasm. This is true for the younger as well as the older woman.

Resolution Phase. The first resolution-phase reaction is the loss of the sex tension flush if it has occurred. Shortly thereafter, occasionally occurring simultaneously, is detumescence of the areolae. This reaction progresses rapidly because of the limited extent of the areolar tumescent reaction. Most of the women in the 50-, 60-, and 70-year age groups retain obvious nipple erection for a matter of hours after an orgasmic experience. Loss of nipple erection after orgasmic release of sexual tension is usually a much more delayed process in postmenopausal as compared to the premenopausal years. However, postorgasmic nipple erection may be an indication of continuing sexual interest following insufficient orgasmic tension release. This distress occurs in the sexually responding woman of any age. Therefore, definition of such etiology for retained nipple erection only can be determined from direct interrogation of the individual woman involved.

As a general clinical observation, it may be stated that the more pendulous and slack the breasts of women of any age, the more resistant the breasts are to the vasocongestive size increase of sexual excitement. This observation has particular application to the postmenopausal woman. As a direct result of the aging process, a significant degree of elasticity is lost from breast tissues, causing sagging and flattening. As hormone levels fall there usually is measurable loss in integral breast tissue and actual breast dimensions. Therefore, reduction in or absence of vasocongestive response in the sagging, flattened breast becomes even more evident with the passing years.

SEX FLUSH

The superficial vasocongestive skin response to increasing sexual tensions develops in approximately 75 percent of women under the age of 40. It does not occur so frequently in older women. Only 14 of the 27 members of the 41–50-year age group showed the mottled maculopapular type of erythematous rash which first appears over the epigastrium late in the excitement phase or shortly after plateau phase has been achieved. Only 3 of the 23 members of the 51–60-year age group evidenced the sex flush at any time during their evaluation, and none of the women past the age of 60 years demonstrated the flush.

When the flush appeared, it spread in the normal fashion over the breasts, appearing first on the anterior and superior breast surfaces and then on the anterior chest wall. In most of the women the flush continued to spread over the shoulders, neck, face, and forehead. In only one individual (41–50-year age group) was the sex flush of sufficient degree to be noted over the back, abdomen, and extremities. In short, the development of the sex flush in the aging female is limited in occurrence and is restricted to the epigastrium, anterior chest, neck, face, and forehead, as opposed to the rather protean distribution seen in the younger woman (see Chapter 3).

MYOTONIA

General muscle-tension elevation in response to sexual stimuli decreases as the woman ages. There obviously is less tension created during voluntary muscle contraction, and specific examples of involuntary striated-muscle spasm, such as carpopedal spasm, are quite rare. The exception to this general rule is created by a woman of 60 or 70 years responding to sexual stimuli as part of a regularly recurring opportunity of exposure to sexual episodes.

URETHRA AND URINARY BLADDER

As in younger women, there is a minimal involuntary distention of the external urinary meatus during an intense orgasm experienced by older women. Menopausal and postmenopausal women have been observed through many cycles of sexual response during which the actual orgasmic phase was of moderate or minimal intensity. In these situations gaping of the urinary meatus usually did not occur. However, when the orgasmic experience was of high intensity or the woman moved from one orgasmic experience to a second or even a third in rapid succession, gaping of the urinary meatus was observed frequently.

Many postmenopausal women complain of burning on urination within the first few hours after coition, particularly if coital connection is continued for extended lengths of time. This clinical distress, identified in younger women as "brides' cystitis," develops from mechanical irritation of the urethra and the bladder produced by the normal thrusting movement of the penis. As the woman moves through her postmenopausal years the lining of the vagina becomes very thin and atrophic. Instead of having the thick, rugal pattern of the hormonally well-stimulated premenopausal vagina, the walls of the postmenopausal vaginal barrel are tissue-paper-thin and, therefore, cannot protect the subjacent structures of the urethra and bladder by absorbing the mechanical irritation of active coition. Therefore, irritation of the urethra and bladder occurs with some regularity and may do so with a high degree of frequency if the aging female does not lubricate well. It is not unusual, then, to find many older women having to contend with a sense of urinary urgency shortly after coital connection and being forced frequently to urinate immediately after coition. Some of these individuals even may complain of urinary burning and frequency for as long as two or three days after an episode of extended coital connection.

One woman aged 57 described rare occasions of involuntary loss of urine during coition with a particularly forceful male partner. The same individual also loses urinary control with coughing and sneezing. She has clinical evidence of both a cystourethrocele and rectocele. She has been catheterized on two occasions immediately after voiding and demonstrated retention of 75–90 cc. of residual urine in the bladder. This instance of coitally connected urinary loss is a reported fact. Urinary incontinence has not been observed in the research laboratory.

THE RECTUM

Contraction of the rectal sphincter during orgasm is generally an indication of the intensity of the specific orgasmic response. Regularly recurring rectal contractions usually are seen in younger women during episodes of multiorgasmic experiences. Suggestive of the possibility of a generalized reduction in the intensity of orgasmic expression as a part of the aging process is the fact that orgasmic-phase rectal contractions have been observed only three times in women beyond the age of 51 years. In each instance the orgasmic phases obviously were associated with severe tension levels, and in two of the three instances immediate return to a second orgasmic experience was anticipated and executed.

THE EXTERNAL GENITALIA

The clitoris and the minor and major labia vary in responsiveness to sexual tensions as the human female ages. Clitoral response continues into the 70-year age groups in patterns similar to those established for the premenopausal female. On the other hand, reactions of the minor and major labia reflect involutionary changes that appear to be inherent in the aging process.

THE CLITORIS

All of the 61 women past the age of 40 who have cooperated in the research effort demonstrated the usual clitoral response patterns of younger women (see Part 1 of Chapter 5). It should be reemphasized that for all ages there is normally marked variation in the anatomic structure of the clitoral body and glans. Clitoral glandes measure 3–4 mm. to 1 cm. in transverse diameter, and both measurements must be considered within the normal limits of anatomic structuring. The rapidity of clitoral reaction to sexual stimulation depends upon whether there is direct manipulation of the mons area or the sexually stimulative activities are focused on other erotic areas of the body (see Part 2 of Chapter 5). If stimulation of the clitoris is other than by direct area contact, there is distinct delay in reaction time as opposed to the speed with which the clitoris reacts to direct stimulation.

Excitement Phase.   Clinically obvious tumescence of the clitoral glans was observed during excitement-phase levels of sexual response in only 14 of the 61 women past 40 years of age. In younger women, approximately 25 of these 61 women would have been expected to demonstrate an obvious tumescence of the clitoral glans. Of the 14 aging women who did show clitoral glans tumescence, 9 were in the 41–50-year age group, 4 were in the 51–60-year age group, and one woman was 67 years of age.

The response of vasocongestive increase in clitoral-shaft diameter which has been established as a reactive constant in premenopausal age groups also was a constant factor in physiologic response of menopausal and postmenopausal women. As sexual tensions rise through excitement toward plateau-phase levels of response, the shaft of the clitoris thickens, providing a diameter increase that in some women exceeds twofold.

Plateau Phase.   As plateau-phase levels of sexual tension are established and the aging female approaches orgasmic-phase release, the clitoris elevates away from its pudendal-overhang positioning, retracting the exposed glans beneath its minor-labial hood in the manner described for younger women (see Part 1 of Chapter 5). This retraction of the clitoral shaft and glans and flattening of the entire shaft on the anterior border of the symphysis (a constant plateau-phase response of younger women) continue unabated as the human female ages. The retraction reaction, when completed, reduces clitoral-body length by approximately 50 percent in the immediate preorgasmic period.

Orgasmic Phase.   There is no established orgasmic-phase reaction of the clitoris, regardless of age of the human female.

Resolution Phase.   Clitoral-body retraction is terminated with extreme rapidity. Lengthening of the shaft returns the clitoris to its normal pudendal-overhang positioning immediately after orgasmic experience. In those few instances in which obvious tumescence of the clitoral glans was established, tumescence was lost within a few seconds after the aging female's orgasmic experience.

THE MAJOR LABIA

The flattening, separation, and elevation of the major labia that develop in response to elevated sexual tensions, particularly in the nulliparous woman, are lost as the woman ages. This reaction normally separates and elevates the labia in an upward and outward direction away from the vaginal outlet (see Chapter 4). Only three women in the 41–50-year age group demonstrated the major-labia elevation reaction. None of the women past the age of 51 showed this response to excitement-phase or even plateau-phase levels of sexual tensions.

The major labia lose fatty-tissue deposits as the reduced hormone levels of the postmenopausal years affect female anatomy. With the loss of major-labial body content also goes some loss of elastic tissue. Therefore, it was not unexpected that the major-labial elevation reaction would be basically altered in the advanced years.

THE MINOR LABIA

The minor labia of younger women undergo a vasocongestive thickening during advanced excitement-phase levels of sexual response which extends the vaginal barrel by approximately 1 cm. This vasocongestive reaction is reduced when the human female ages. Minor labial thickening and expansion still was obvious in 18 of the 27 members of the 41–50-year age group, and in 7 of the 23 members of the 51–60-year age group. None of the 11 women past the age of 61 years evidenced this vasocongestive reaction.

The minor-labia reaction of younger women, specific to the plateau phase of the sexual cycle, is a definitive color change that ranges from a cardinal-red to a burgundy-wine color and occurs in the immediate preorgasmic phase of the sexual response cycle. This sex-skin reaction of the minor labia is pathognomonic of impending orgasm in the premenopausal human female (see Chapter 4).

As the human female ages, there is an obvious loss in the consistency of the minor-labial sex-skin reaction. All the women in the 41–50-year age group demonstrated the sex tension color change immediately prior to orgasmic experience. Nineteen of the 23 members of the 51–60-year age group also underwent the minor-labial color change immediately prior to orgasmic experience. Only 2 of the 8 members of the 61–70-year age group demonstrated the color change, and only one of the 3 members of the 71–80-year age group still retained this vasocongestive responsiveness of the minor labia. These elderly women are the only women observed through orgasm who did not demonstrate the preorgasmic color change of the minor labia.

BARTHOLIN'S GLANDS

The secretory activity of Bartholin's glands is somewhat slowed by the aging process, but not until the human female is well into the postmenopausal years. It may be recalled that Bartholin's-gland secretory activity in younger women develops only during the plateau phase of the sexual response cycle (see Chapter 4). Even then it is produced only if plateau-phase tension levels are maintained for extended periods of time or if coital connection purposely is continued for many minutes. Normally, the amount of secretory material is very small, usually a drop or two of the mucoid substance, and its only known use is to lubricate the vaginal outlet during long-maintained coital connection.

All 27 members of the 41–50-year age group demonstrated Bartholin's gland secretory activity during the plateau phase of at least one of their observed cycles of sexual response. However, such secretory activity was present in only 12 of the 51–60-year age group, in 3 of the 61–70-year age group, and was not observed in the 71–80-year age group. Since Bartholin's gland activity is extremely difficult to demonstrate, it may well be that there were cycles of sexual response in which individuals did produce this mucoid material and the secretory activity was overlooked. In older women there not only was a marked reduction in demonstrable secretory activity, but the amount of material produced also was significantly reduced when compared to that developed by younger women.

THE REPRODUCTIVE VISCERA

THE VAGINA

The aging woman's vagina undergoes specific involutionary changes which should be described in some detail before attempting to establish the differences in vaginal response to sexual tensions between younger and older women. After the woman has undergone the normal menopausal involution of ovarian sex-steroid production, changes develop in the target organs, i.e., the labia, vagina, uterus, breasts, etc. The well-stimulated healthy vagina of the 30-year-old woman has an entirely different appearance from that of the steroid-starved woman in the 61–70-year age group. After the ovaries cease or grossly reduce sex-steroid production, the walls of the vaginal barrel begin to involute. Instead of having the well-corrugated, thickened, reddish-purple appearance of the well-stimulated vagina, the walls of the senile vaginal barrel become tissue-paper-thin, lose the rough, corrugated look, and change to a light pinkish color. The very thin walls of the senile vagina almost give the impression that they can be seen through.

In addition to a thinning of the mucosa with aging, there is shortening of both vaginal length and width (at the transcervical level). The vaginas of the 11 women past the age of 60 who co-operated with the research program measured 4.5–6 cm. in length and 1–1.5 cm. in width (transcervical level) of the vaginal barrel, as opposed to a measurement of 7–8 cm. in length and approximately 2 cm. in width established previously for normally menstruating women (see Part 1 of Chapter 6).

In addition to loss of length and width during the aging process, the vagina also loses some of its expansive ability, as might be anticipated from the reported loss of vaginal-wall thickness. The involuntary neuromuscular response to sexual tensions which results in expansion in vaginal length and in transcervical width obviously is influenced by states of sex-steroid starvation. One woman, aged 62, has been with the experimental program for nine years. No sex-steroid replacement therapy has been administered despite several years of steroid-withdrawal symptomatology.

The vaginal barrel has lost length, transcervical width, and a significant degree of involuntary ability to expand under sex tension influence.

Excitement Phase. As excitement-phase levels of sexual tension are achieved, the first evidence of physiologic response is, of course, the production of vaginal lubrication. This primary evidence of female sexual tension is affected significantly by advancing years. Once the individual female is approximately five years past the cessation of her menses, the rate and the amount of lubrication production diminish to an obvious degree. This is a general rather than a specific statement of fact, for there have been and are individual exceptions to this rule.

In younger women, vaginal lubrication is well distributed throughout the vaginal barrel within 10–30 seconds of the onset of any form of effective sexual stimulation. Once the individual female is beyond the midfifties, and particularly when she is beyond 60 years of age, it may take from one to three minutes before any definitive production of vaginal lubrication can be observed, despite the fact that the woman obviously is responding with real anticipation and pleasure to the particular form of sexual stimulation employed.

Three women represent the observed exceptions to the rule of delayed lubrication production for the aging female—two in the 61–70-year age group, and one of 73 years. All three women consistently respond to sexual stimulation with rapid production of vaginal lubrication in a manner expected from a 20–30-year-old woman. For these three women lubrication diffuses throughout the vaginal barrel and covers the minor labia in short order. This rapid, full production of lubrication occurs despite the fact that in all three instances the vaginal mucosa is very thin and atrophic. The only possible explanation for these exceptions to the general rule of slowed lubrication production with aging is the interesting fact that these three women (two in the 60- and one in the 70-year age group) have maintained active sexual connections once or twice a week throughout their mature lives. They are the only ones in the over-60-years age groups to have maintained coital connection at such a frequency level.

Excitement-phase involuntary expansion of the inner two-thirds of the vaginal barrel is reduced as has been described, in degree and in rapidity of reaction during the postmenopausal years. However, the vagina does respond to the direct stimulation of actual mounting opportunities with an expansive ability in excess of that demonstrated during orgasmic cycles induced by manipulation. As opposed to this older-age response pattern, the younger female expands the inner two-thirds of the vagina almost as well with manipulative activity as she does during active coition.

Plateau Phase. Since the inner two-thirds of the vaginal barrel expands more slowly as the woman ages, this reaction may be observed as frequently at plateau levels of sexual response as during the excitement phase. This is in opposition to younger women's reactions. Most vaginal expansion in length and in transcervical width has been accomplished before plateau-phase levels of sexual tension have accumulated.

The major physiologic response to plateau-phase levels of sexual tension is the development of the orgasmic platform in the outer third of the vagina (see Part 1 of Chapter 6). This reaction occurs in all women at all ages and subsequent to any and all forms of effective sexual stimulation. Local vasocongestion in the outer third of the vagina is reduced significantly in intensity after senile involution of the vaginal walls and constriction of the vaginal barrel have developed. One woman aged 62, with nine years of cooperation with the program, now develops an orgasmic platform that is approximately one-half as extensive a reaction as the platform produced during her middle fifties.

Once the orgasmic platform is developed fully, the central lumen of the senile vagina is constricted to a degree proportional to that of younger women. This is a constant reaction despite the reduced local vasocongestion. The marked constriction of the vaginal lumen probably is due to the fact that there is reduction of the total vaginal-barrel volume and its involuntary expansive qualities with senile involution of the vaginal walls and surrounding tissues.

Orgasmic Phase. The characteristic physiologic reaction to orgasmic levels of sexual tension is contraction of the orgasmic platform. In older women contractions develop in fashion identical to those of younger women with the exception that the orgasmic phase generally is reduced in duration when compared to that of 20–30-year-old women. Postmenopausal orgasmic-platform contractions usually recur from 3 to 5 times, as opposed to the response pattern of younger women, whose orgasmic platform contractions recur normally from 5 to 10 times (see Part 1 of Chapter 6). There are exceptions to the general rule, as evidenced by the three women mentioned previously, two in their sixties and one in her seventies, who have maintained regularly recurring coital connections during their entire mature lives. These three women have orgasmic-platform contractions that have been observed to recur from 4 to 6 or even 7 times, as opposed to the 3 or 4 platform contractions of women of the same age whose opportunity for sexual expression has been delimited by physical or social circumstances.

As is true for younger women, contractions of the orgasmic platform develop in the older female regardless of whether clitoral area manipulation, active coition, or stimulation of any other erotic area is used to develop orgasmic-phase response.

Resolution Phase. The expanded inner two-thirds of the vagina shrinks back to a collapsed unstimulated state with marked rapidity. This is a rapid involution of the entire vaginal barrel rather than the irregular zonal type of reaction that slowly drops the cervix of the anteriorly placed uterus into the transcervical depth of the vagina in the younger woman. This rapidity of vaginal-wall collapse in older women as opposed to the slower involution of the younger woman well may be the result of the increasing rigidity and lack of elasticity in the senile vaginal barrel. The orgasmic-platform vasocongestion is lost even more rapidly than in younger women. This rapid rate of involution again may result from generalized reduction in the extent of pelvic vasocongestive response to sexual tension.

THE CERVIX

As is true for younger women, there never has been evidence of cervical secretory activity in postmenopausal women during any of the four phases of the sexual response cycle.

During resolution, a slight patulousness of the nulliparous external cervical os has been demonstrated frequently in the younger female. This dilatation of the external cervical os never has been observed in any woman who is more than five years past the cessation of menses.

THE UTERUS

As the human female experiences endocrine starvation during her involutionary years, the cervix and the uterus respond to the deprivation of sex-steroid stimulation by shrinking in size, with the greatest evidence of involutionary change in the corpus, or body, of the uterus. If the uterus is anteriorly placed, there is some elevation of the senile corpus as excitement and plateau phases develop in the sexually responding older woman. Thus, a minor tenting effect develops at the transcervical depth of the vaginal barrel. Uterine elevation is not as marked as that seen in younger women. The tenting effect in the woman now 62 years of age who has been with the program for nine years is much less marked in extent, and the degree of uterine elevation is reduced significantly when compared to the severity of these reactions when she first joined the program. Uterine elevation, if it is to occur in older women, develops either in advanced excitement or any time during a plateau-phase of sexual tension. This is further evidence of aged women's delayed reaction time when compared to the usual pattern for the younger age groups. With premenopausal women, uterine elevation essentially is completed by the time plateau-phase levels of sexual tension have been established. No evidence of vasocongestive uterine enlargement has developed in any of the postmenopausal women.

As senile pelvic involution progresses after ovarian-steroid production is no longer adequate for target-organ protection, the uterus shrinks in size so that when the average woman is five to ten years past cessation of flow, the uterus and cervix are essentially equal in length. For this reason it has been impossible to place intrauterine electrodes successfully, and orgasmic-phase contractility of the senile uterus has not been recorded, as it has been for younger women (see Chapter 8). However, several of the women in the 60–70-year age group have responded to the stimulation of orgasm with the clinical suggestion of uterine con-tractility reported as severe cramping pain. One individual in th 60-year age group describes the uterine contractions of orgasm a "almost like labor pains except that they occur more rapidly. There seems to be little doubt that the factor of uterine contractilit with orgasm remains in senile women. The severity, the duratio and the degree of recurrence of these contractions obviously vari tremendously from individual to individual and within the sam individual depending upon the intensity of the orgasm. There no definitive information available at present as to the physiolog response of the senile uterus to effective sexual stimulation.

In brief, significant sexual capacity and effective sexual pe formance are not confined to the human female's premenopausa years. Generally, the intensity of physiologic reaction and duratio of anatomic response to effective sexual stimulation are reduce through all four phases of the sexual cycle with the advancin years. Senile involution of the target organs (breasts, labia, vagin. uterus) is evidence of postmenopausal states of sex-steroid starv. tion. Regardless of involutional changes in the reproductive organ the aging human female is fully capable of sexual performanc at orgasmic response levels, particularly if she is exposed to regula ity of effective sexual stimulation. Steroid starvation has the pr mary influence of reducing rapidity and intensity of physiolog response. When reduction in psychologic tension levels develops usually is secondary to considered loss of physiologic capacity an not a direct effect of steroid starvation.

2. CLINICAL CONSIDERATIONS

Theoretical knowledge and clinical experience related to sexu problems of the aging are totally inadequate to meet the requir ments of men and women who currently are living within th framework of our newfound longevity. Any counselor facing pro lems created by the sexual tensions of menopausal or postmen pausal women finds himself seriously handicapped by the lack of a well-established body of literature on the subject.

Reports of the aging female's sexual activity have been limited largely to studies of the menopausal or immediate postmenopausal years. Possibly this investigative concentration on the climacteric age of 45 to 55 years has been stimulated by women's tendency to seek relief at this time from a variety of psychophysiologic problems. In order to establish the aging-female component of the study-subject population, 157 intake interviews were conducted with women beyond 51 years of age. One hundred fifty-two of these women contributed detailed sociosexual histories in response to team interrogation. From this material, together with that accumulated from seven years of clinical therapy of sexual inadequacy, the behavioral concepts expressed in this chapter have been drawn. Only 34 of the original 157 women interviewed cooperated actively in the investigative program (see Part 1 of this chapter). The age distribution of the 152 women past 51 years of age and the level of their formal education are listed in Table 15-2. The subjects provided histories separately to both the male and female members of the interview team.

The degree of influence of sex-steroid withdrawal upon female sexual adjustment during the menopausal and postmenopausal years has not been established, although it is a popular practice to assign to the physiologic fact of steroid starvation most of the physical ills and psychosexual problems associated with these

TABLE 15-2

Age and Education of 152 Geriatric Female Study Subjects

Age by Decade	No. Subjects	Education			
		Grade School	High School	College	Graduate School
51–60	98	2	62	29	5
61–70	37	1	28	6	2
71–80	17	3	11	3	0
Totals	152	6	101	38	7

years. Many facets of the relationship between states of steroid starvation and female sexual response remain to be defined.

There are several mechanical factors occasioned by endocrine imbalance which result indirectly in painful coition during the postmenopausal years. Many women who have never been discomforted by sexual activity complain of physical distress during or shortly after coital connection in the immediate postmenopausal years. Coition may become severely painful during the penetration phase, or extended coital connection may be followed by vaginal burning, pelvic aching, or vague lower abdominal distress. Frequently, coition is followed by burning and irritation on urination. These symptoms of dyspareunia and dysuria may continue for 24 to 36 hours after sexual connection.

The symptoms of acquired dyspareunia and dysuria usually result from a marked thinning of the vaginal mucosa and a reduction in involuntary distensibility of the entire vaginal barrel. The natural ability to lubricate the vaginal barrel and introitus effectively may be reduced or the reaction time slowed for women beyond their middle fifties (see Part 1 of this chapter).

Thinning of the vaginal walls, reduction in length and transcervical diameter of the vaginal barrel, and shrinking of the major labia, leading to constriction of the vaginal outlet, result from sex-steroid starvation as ovarian function fails. These specific indications of postmenopausal physiologic involution of ovarian function may be corrected easily with adequate endocrine-replacement therapy [52, 80, 86, 200, 201]. The return of physical capacity for effective sexual performance should be considered the indirect result of removing the physical roadblocks of target-organ (vagina) senility, rather than the primary result of direct hormone stimulation of lagging sexual tensions.

Even more necessary for maintained sexual capacity and effective sexual performance is the opportunity for regularity of sexual expression. For the aging woman, much more than for her younger counterpart, such opportunity has a significant influence upon her sexual performance. Three women past 60 years were repeatedly observed to expand and lubricate the vagina effectively despite obvious senile thinning of the vaginal walls and shrinking of the major labia. These women have maintained regular coital connection once or twice a week for their entire adult lives.

Frequently, women from five to ten years postmenses who experience infrequent coition (once a month or less) and who do not masturbate with regularity have difficulty in accommodating the penis during their rare exposures to coition. It also is true that many younger women deprived of coital opportunity for long periods of time may have to contend with a slowed rate of vaginal lubrication and restricted vaginal-barrel expansion during a first return to coital connection. However, their difficulties are far less pronounced than those of older women in similar circumstances of coital deprivation, and their full physiologic response to coital stimulation is established far more rapidly.

There is another manifestation of steroid imbalance in the sexual response patterns of the aging human female. As women age and lose their sex-steroid levels, uterine contractions occurring with orgasm frequently become painful. The actual degree of distress varies from time to time and from woman to woman, but when experienced, this painful uterine cramping develops during as well as subsequent to orgasmic expression. While these uterine contractions occur in women of all ages experiencing orgasmic response, younger women rarely have accompanying physical discomfort that reaches a level of clinical distress.

Beyond 60 years of age some women are so distressed with these contractions that they purposely avoid orgasmic experience and even coital connection if possible. The pain from the contractions is relieved by combinations of estrogen and progesterone, if both are supplied continuously in a balanced combination to the distressed postmenopausal woman. Neither hormone used singly will relieve severe degrees of uterine contractile distress. It should be emphasized that only pain of the contractions is lost by adequate hormone replacement. The uterine contractions continue to occur regularly with orgasmic expression.

Thus the simple fact remains that if opportunity for regularity of coital exposure is created or maintained, the elderly woman suffering from all of the vaginal stigmas of sex-steroid starvation still will retain a far higher capacity for sexual performance than her female counterpart who does not have similar coital opportunities.

As has been seen, endocrine starvation has an indirect influence upon, but certainly not absolute control over, female sexual capacity or performance. Steroid starvation also has an indirect influence upon female sexual drive. However, sex drive is but one in the total of physical and psychosocial factors influenced by the aging process.

It has become increasingly evident that the psyche plays a part at least equal to, if not greater than, that of an unbalanced endocrine system in determining the sex drive of women during the postmenopausal period of their lives. If endocrine factors alone were responsible for sexual behavior in postmenopausal women (whether menopause occurs by surgical or natural means), there should be a relatively uniform response to the physiologic diminution and ultimate withdrawal of the sex hormones. However, there is no established reaction pattern to sex-steroid withdrawal. For instance, clinical symptoms of menopausal distress vary tremendously between individuals, and, for that matter, within the same individual as the demand arises for increased physical or mental activity [204].

Elevation of sexual responsiveness rarely results directly from the administration of estrogen or estrogen-like products. Estrogenic compounds frequently do improve sex drive in an indirect contribution above and beyond the original intended purpose of insuring a positive protein balance in the aging female. A woman previously experiencing a healthy libido may become relatively asexual while contending with such menopausal discomforts as excessive fatigue, flushing, nervousness, emotional irritability, occipital headaches, or vague pelvic pain. This individual's personal eroticism may be restored to previously established response levels following the administration of estrogenic preparations. The obviously increased sex drive well may have developed secondary to relief of the woman's multiple menopausal complaints, rather than as a primary or direct result of the actual adjustment of the individual's sex-steroid imbalance [201].

Personality studies of menopausal or postmenopausal women are more prevalent in the literature than are endocrine studies.

In the opinion of Stern and Prados [302] there is no correlation between the intensity of the type of physical symptoms usually related to hormonal withdrawal (hot flushes, for instance) and the severity of emotional disturbances occasioned by steroid starvation. They do feel, however, that among the many complaints presented by menopausal women the physical complaint of pelvic pain is the most intimately associated with the more severe forms of psychic maladjustment. Rosenzweig [270] suggests that emotional disturbance during the menopause may represent a reaction of frustration to the representation of the menopause as the failure of the whole life cycle in respect to procreation. Shorr [293] came to the conclusion that the emotional complications of the menopause are basically psychoneurotic in nature and are almost always exacerbations of similar disorder patterns developed earlier in the patient's life. Certainly, absence of a sense of well-being and general physical discomfort frequently present in the menopausal woman only would tend to heighten and reactivate established psychoneurotic behavior patterns of sexual origin. The average woman's psychosomatic symptoms fluctuate to the greatest extent in the menopausal years. It is to be expected that the sex drive, with its multiple related tensions, would reflect the instability of this age group.

During the climacteric either a return to or escape from the reproductive drive has been demonstrated by many women. Helene Deutsch [53] has stated that during the preclimacterium many women develop an overwhelming desire to become pregnant once more, demonstrating in this manner an apprehensive feeling about the "closing of the gates." Other women welcome the advent of the climacteric with genuine pleasure but do not demonstrate an increased sex drive until menopause obviously is well established. These women usually have been burdened with either an excessive number of children or a financial situation too insecure to guarantee adequate family protection. They develop a resultant "freedom from fear of pregnancy" as the menses terminate.

Many a woman develops renewed interest in her husband and in the physical maintenance of her own person, and has described a "second honeymoon" during her early fifties. This expression of unleashed sexual drive occasioned by the alleviation of the "pregnancy phobia" is one of the most frequently occurring factors responsible for increased sexual tensions evident in the 50–60-year age group. Noteworthy is the obvious fact that the renewed husband-interest of the pregnancy-phobic individual reflects a baseline of pleasure and stability in the sexual relationship.

When the women who demonstrate the "freedom from fear" complex are added to those concerned with the "closing of the gates," the frequently increased levels of sexual activity during the late forties and early fifties noted by many observers is partially explained. It should be emphasized, however, that the woman who increases her sexual activity basically from a desire to conceive rarely has major interest in the sexual relationship per se. Thus, the marked increase in sexual activity of these two groups does not reflect parallel increase in sex drive.

Absolute contraceptive security has not been available in the past to women who are presently in the menopausal and post-menopausal age groups. Therefore, the pregnancy phobias, when they have developed, have been thoroughly understandable. When the 20–30-year-old women of today are in their late forties and early fifties, the expected increase in sex drive concomitant with release of pregnancy fears well may be a thing of the past. In today's society the young wife need have no fear of unwanted pregnancy, provided her religion tolerates the practice of contraception and she can afford to purchase the effective contraceptive materials presently available.

It also should be recalled that women beyond 50 years of age usually have resolved most of the problems associated with the raising of a family. Once the exhausting physical and extensive mental demands of brood protection have been obviated by the maturing of the family group, it is only natural that new directions are sought as outlets for unexpended physical energy and re-awakened mental activity. Thus, a significant increase in sexual activity marks the revived sex drive of these middle-aged women. Frequently this is the time for casting about for new sexual partners or for the development of variations of or replacements for long-established unsatisfying sexual practices.

The Kinsey group has noted that a large part of the sex drive during the postmenopausal age is related directly to the sexual

habits established during the procreative years [144]. The interview material suggests that a woman who has had a happy,
well-adjusted, and stimulating marriage may progress through the
menopausal and postmenopausal years with little or no interruption in the frequency of, or interest in, sexual activity. Additionally,
social and economic security are major factors in many women's
successful sexual adjustment to their declining years.

Needless to say, there is an increasingly large segment of the
female population that is diametrically opposite to the reasonably
adjusted individual described above. If a woman has been plagued
by seeming frigidity, or by lack of regularly recurrent or psycho-
sexually satisfactory coital activity during her active reproductive
years, there is reason to believe that the advent of the postmenopausal years may serve to decrease sex drive and to make the
idea of any form of sexual expression increasingly repugnant. This
individual uses the excuse of her advancing years to avoid the
personal embarrassment of inadequate sexual performance or the
frustrations of unresolved sexual tensions.

There also remains the Victorian concept that older women
should have no innate interest in any form of sexual activity. The
idea that the postmenopausal woman normally should have little
or no sex drive probably has arisen from the same source. Even
dreams or fantasies with sexual content are rejected in the widespread popular belief that sexual intercourse is an unsuitable
indulgence for any woman of or beyond middle age.

As emphasized by Newman and Nicols [240], the sexual activity
of the woman in the 70-plus age group unfortunately is influenced
by the factor of male attrition. When available, the male marital
partner is an average of four years older than the female partner.
Many of the older husbands in this age group are suffering from
the multiple physical disabilities of advancing senescence which
make sexual activity for these men either unattractive or impossible.
Thus, the wives who well might be interested in some regularity
of heterosexual expression are denied this opportunity due to their
partner's physical infirmities. It also is obvious that extramarital
sexual partners essentially are unavailable to the women in this
age group.

The trend of our population toward an aging society of women without men must be considered. Roughly 10 percent of women never marry. In addition, the gift of longevity has not been divided equally between the sexes. As a result, there is a steadily increasing legion of women who are spending their last years without marital partners [222]. Many members of this group demonstrate their basic insecurity by casting themselves unreservedly into their religion, the business world, volunteer social work, or overzealous mothering of their maturing children or grandchildren. Deprived of normal sexual outlets, they exhaust themselves physically in conscious or unconscious effort to dissipate their accumulated and frequently unrecognized sexual tensions.

Masturbation presents no significant problem for the older-age-group women [205, 215]. The unmarried female who has employed this method for relief of sexual tensions during her twenties and thirties usually continues the same behavioral pattern during her forties and through her sixties. When heterosexual contacts are limited or unavailable the widowed or divorced woman also may revert to the masturbatory practices of her teens and twenties when sexual tensions become intolerable. As might be expected, there is reduction in the frequency with which manipulative relief is deemed necessary beyond 60 years of age.

There seems to be no physiologic reason why the frequency of sexual expression found satisfactory for the younger woman should not be carried over into the postmenopausal years. The frequency of sexual intercourse or manipulative activity during the post-menopausal years is of little import, as long as the individuals concerned are healthy, active, well-adjusted members of society.

It would seem that the maladjustments and abnormalities of sex drive shown by states of hyper- or hyposexuality which develop during and after the menopause might best be treated by prophylaxis. If satisfactory counseling of sexual content were made more available to sexually insecure, uneducated, or inadequate women in the premenopausal years, there is reason to believe that the unresolved tensions of the later years might be reduced or, to a large extent, avoided. There is no reason why the milestone of the menopause should be expected to blunt the human female's sexual capacity, performance, or drive. The healthy aging woman normally has sex drives that demand resolution. The depths of her sexual capacity and the effectiveness of her sexual performance, as well as her personal eroticism, are influenced indirectly by all of the psycho- and sociophysiologic problems of her aging process. In short, there is no time limit drawn by the advancing years to female sexuality.

16

THE AGING MALE

1. ANATOMY AND PHYSIOLOGY

The aging male's anatomic and physiologic responses to effective sexual stimulation have been investigated during the past decade. This discussion of male sexual capacity and performance is based on data obtained from 39 men whose ages ranged from 51 to 89 years at the time of their evaluation (Table 16-1).

As was true for their aging female counterparts, it was extremely difficult to elicit active cooperation from even this small group of men. The material to be presented must be accepted in the light of an admittedly inadequate study-subject population. This is particularly true for results reported from men over 70 years of age. However, the returns from this limited number of aging males provide opportunity for comparison with patterns of sexual response firmly established for younger men and permit superficial consideration of the effect of the aging process on male sexual physiology. Men in the 51–60-year age group were included in the discussion to provide a parallel to the menopausal women's response patterns. The four phases of the human cycle of sexual response will serve as a means of descriptive comparison.

Sexual response patterns were described not only by the active study subjects but also by the 212 men beyond 50 years of age who cooperated with sociosexual interviews (see Part 2 of this chapter).

As the male ages, the major differences in sexual response relate to the duration of each of the phases of the sexual cycle. As opposed to the younger man's well-established reaction pattern of immediate erection, early mounting, and rapid ejaculation, the older man (particularly over 60 years old) is slower to erect, to mount, and to ejaculate. The resolution-phase refractory period also lengthens for the male past the age of 50 years.

TABLE 16-1

Age Distribution of 39 Geriatric Male Study Subjects

Age Distribution	No. Active Participants
51–60	19
61–70	14
71–80	4
81–90 *	2
Total	39

* Oldest study subject was 89 years old.

If there has been a well-adjusted marital pattern of frequency of coital exposure, it usually is maintained well into the fifties by healthy males. As a rule, the urban male slows in tension increment before his rural counterpart. Assuming equally good health and coital opportunity, this may be more a reflection of psychosocial distraction than of specific physiologic involution.

With rare exceptions the male over 60 years old usually will be satisfied completely with one or, at the most, two ejaculations a week regardless of the number of coital opportunities or the depth of his female partner's sexual demand. Many men in their middle or late fifties and in their sixties find that they cannot redevelop penile erection for a matter of 12 to 24 hours after ejaculation. Those who achieve a relatively early return to erection may have lost their ejaculatory urge and are perfectly content to serve their female partners to the completion of the woman's sexual demands without recurrent ejaculatory interest.

EXTRAGENITAL REACTIONS

THE BREASTS

As is true for younger men, there is only one general anatomic reaction of the male breast to effective sexual stimulation—nipple erection during the plateau phase of sexual response. Few men under 60 years of age ejaculate without an obvious turgidity, if not full erection, of the nipples. As the male ages, however, the degree of nipple turgidity is reduced. Four men in the 61–70 age group, 3 in the group between 71 and 80 years, and both men over 80 years of age showed no clinically discernible nipple erection. Whether this lack of nipple sensitivity is evidence of loss of elastic-tissue substrate in the nipples or reduction in the intensity of the body's physiologic responses to orgasm, or both, cannot be determined at present.

Of interest is the fact that loss of nipple erection in the resolution phase usually is delayed in the aging male. On occasion, men 60 years or older have been noted to maintain nipple erection for hours after ejaculation, unless the opportunity to sleep has intervened.

THE SEX FLUSH

Following the aging human female's reaction pattern, the aging male loses ability to develop the vasocongestive maculopapular flush of sexual tension. Only 2 of the 39 males past the age of 50 years were observed to develop the sex flush. One of these men demonstrated a fleeting measles-like rash confined to the epigastrium and anterior chest wall, whereas the other man developed a full-blown sex flush over the diaphragm, anterior chest, neck, face, and forehead, but not on the back or extremities. In younger males, the sex flush (plateau phase-oriented in its timing) usually is evidence of severe levels of sexual tension. When an older man developed the sex flush prior to ejaculation, clinical observation and his subsequent vocalization created the impression that the orgasmic experience was more severe than his average intensity of response.

MYOTONIA

There is little incidence of involuntary muscle spasm such as carpopedal spasm late in plateau or during ejaculation in the male over 60 years old. Regularly recurring contractions of the musculature of the target organs develop with obvious reduction in intensity. Exceptions to this statement are created by men with continued frequency of exposure to sexual episodes.

THE RECTUM

Rectal-sphincter contractions, which in younger men occur regularly during orgasmic experience, decrease in frequency as the male ages. Regularly recurrent contractions of the rectal sphincter have been noted in only 4 males over 50 years of age: 2 were in their fifties, 1 was in his sixties, and 1 man (a single observation) was 74 years of age. This reduction in involuntary rectal-sphincter tension may reflect a generalized reduction in physiologic intensity of orgasmic experience as the human male ages, a suggestion not too difficult to accept in light of further information to be presented.

THE EXTERNAL GENITALIA

THE PENIS

Excitement Phase. Penile erection normally develops with extreme rapidity in young males. The penis may reach full erection from an unstimulated flaccid state within 3 to 5 seconds of the onset of any form of sexual stimulation (see Part 1 of Chapter 12). This reaction time is at least doubled and frequently trebled as the individual male passes through his fifties and into the 60- and 70-year age status. Generally, the older the male is, the longer it takes to achieve full penile erection, regardless of the effectiveness of the stimulative techniques employed.

Once achieved, penile erection in the aging male may be and frequently is maintained for extended periods of time without ejaculation. This degree of ejaculatory control may be acquired by the younger man with specific training or by avoiding great intensity of or marked variation in stimulative techniques. For the aged male such stimulative restraint or specific training usually is not necessary. Regardless of the variety or effectiveness of the sexually stimulative activity, the ability to maintain penile erection over long periods of time without an ejaculatory sequence is associated with the aging process. Whether this ejaculatory control is the result of wide coital experience or truly reflects a reduction in the intensity of sexual response inherent in the aging process has not been determined. The clinical impression persists that both factors probably are involved in the aging male's improved ejaculatory control.

The younger man experiencing excitement-phase levels of sexual tension over an extended period of time may attain a full erection, partially lose it, and fully regain it several times during any sexual cycle. When full penile erection has been attained by the male over 60 years of age and subsequently lost without ejaculation, increasing difficulty may be encountered in returning to full erective performance. Older males may react to loss of penile erection without ejaculation with what might be termed a secondary refractory period. Once erection has been attained and then lost without an ejaculatory opportunity, many older men experience difficulty in returning to excitement-phase levels of physiologic response, regardless of continuation of the previously effective stimulative techniques. This type of reverse refractory period—i.e., occurring during excitement rather than resolution phase of the sexual cycle—rarely occurs in potent males under 50 years of age.

Plateau Phase. Full penile erection frequently is not attained by the aging male (particularly by those over 60) until just before the ejaculatory experience. The younger male may undergo an involuntary congestive increase in the circumference of the glans at the coronal ridge as the ejaculatory episode approaches (see Part 1 of Chapter 12). The aging male not only experiences this involuntary increase in glans circumference occurring late in the plateau phase but also experiences increased length and diameter of the entire penile shaft just prior to ejaculation. The rapid penile erective pattern associated with the younger male's excitement-phase levels of sex tension is transformed by the slower reactive qualities inherent in the aging process to a plateau-phase penile erective potential for the man over 60 years of age. The color change of the glans penis (particularly at the coronal ridge) that occurs in men younger than 40 has not been observed in men after the age of 60 years.

Orgasmic Phase. The aging male's expulsive penile contractions are established by regularly recurring contractions of the sphincter urethrae, the bulbospongiosus, the ischiocavernosus, and the transverse perineal muscles, paralleling the younger man's physiologic response patterns (see Part 1 of Chapter 12). Contractions of these muscles expand and extend the penile urethra, distend the urethral bulb, and force the seminal fluid from the prostatic and membranous portions of the urethra to and through the external urethral meatus. The younger male can expel the seminal fluid the full length of the penile urethra under such pressure as to deposit initial portions of the seminal plasma 12 to 24 inches from the unencumbered urethral meatus. The man over 50 years of age exhibits markedly reduced ejaculatory prowess, 6 to 12 inches being the average distance that the seminal plasma can be expelled. If penile erection has been maintained for an extended period of time, the actual ejaculatory process may be one of seminal-fluid seepage from the external urethral meatus rather than the usual ejaculatory response with the seminal fluid under obvious pressure.

The intercontractile intervals between the first few penile contractions are similar in timing for both younger and older men. These expulsive contractions occur at intervals of 0.8 second. The older the male, the fewer the number of expulsive contractions and, as mentioned previously, the less the severity of expulsive force propelling the seminal fluid.

Although the male over 60 also starts the ejaculatory process with an intercontractile interval of 0.8 second, the ejaculatory contractions which produce significant expulsive force are reduced in number to one or two at the most. The intercontractile interval rapidly lengthens, particularly after the second expulsive contraction of the penile musculature.

Resolution Phase. There are two major differences between the older and the younger man during the refractory period of the resolution phase: (1) The refractory period lasts for extended periods of time as the male ages, particularly after the age of 60; and (2) penile detumescence (immediately after ejaculation) usually is so rapid in the aging male that first and secondary stages of detumescence, so characteristic of younger men (see Part 1 of Chapter 12), cannot be established.

The late vasocongestive response of full penile erection that for older men frequently is not accomplished until an advanced plateau phase has been established is dissipated in such a rapid fashion that the postejaculatory elderly male may have no recognizable staging during penile involution. Seconds after a 60-year-old man ejaculates the penis may have returned to an unstimulated flaccid state.

THE SCROTUM

Excitement Phase. The older man's scrotal integument (skin and tunica dartos) in a sexually unstimulated state is quite thin and characteristically demonstrates marked relaxation, with multiple folding and free movement over the subjacent scrotal content. There is even more relaxation, folding, and sagging of scrotal tissue as the male ages, since a significant degree of scrotal skin elasticity is lost.

As sexual tensions mount in the younger male, the scrotal folding patterns may be obliterated as the result of a notable tensing and thickening of scrotal integument. These responses to sexual tension are accomplished by localized vasocongestion (see Chapter 13). In the man over 60 years of age, however, scrotal vasocongestive response to sexual tensions is reduced markedly. Frequently, there is no evidence of any localized scrotal vasocongestion. The integumental folding pattern usually persists without tensing or flattening, and thickening of the skin and dartos may not occur. This lack of superficial vasocongestive response to sexual tension persists despite the fact that the aging male obviously is responding successfully to sexual stimulation.

When the scrotal sac contracts in the younger male and thus decreases the internal scrotal diameter, support is contributed secondarily to the reaction of testicular elevation. In the older male the reaction of testicular elevation, as it occurs, receives relatively little support from the nonelastic, sagging, usually noncongested scrotal integument. Occasionally a full scrotal vasocongestive reaction develops, but there is no consistency to the performance.

Plateau and Orgasmic Phases. There are no specific scrotal reactions to either plateau- or orgasmic-phase levels of sexual tension in younger or older men.

Resolution Phase. The younger man resolves the engorged scrotal integument in two different patterns. The more frequent reaction is characterized by rapid loss of the congested, tense. appearance of the scrotum and early reappearance of the rugal pattern. However, approximately 25 percent of all young men undergo slow involution from the vasocongested orgasmic state. Full loss of scrotal vasocongestion may be delayed as long as one to two hours after the ejaculatory experience.

Since full vasocongestion of the scrotal integument is a relatively rare occurrence for the 60-year-old male, and, for that matter, for many 50-year-old males no definitive resolution pattern has been established. Suffice it to say that when scrotal vasocongestion in clinically observable degree occurs in response to sexual stimulation in older men, it usually follows a slow involutionary pattern.

As in the younger male, no specific color change of the scrotal integument has been observed in any phase of the sexual response cycle.

THE TESTES

Excitement and Plateau Phases. The reaction of testicular elevation results from a shortening of the spermatic cords and is established by contraction of the cremasteric musculature which invests these cords (see Chapter 13). Testicular elevation occurs in younger males in late excitement phase or early in plateau phase of the sexual cycle. With few exceptions, males under 50 years cannot ejaculate until full testicular elevation has occurred.

In males beyond the midfifties, testicular elevation is reduced in excursion. Frequently, older males have been observed to ejaculate with the testes elevated only one-third or one-half way to the perineum. The right testicle may elevate almost to full perineal apposition, yet the left testicle will not elevate more than a half or two-thirds of the way from the scrotal depths. There is no real significance in older men's plateau-phase as opposed to younger men's excitement-phase timing for the testicular-elevation reaction. This is another example of the older man's delayed reaction time in sexual response.

When testes elevate from the scrotal depths, they do so any time from onset of sexual stimulation to shortly before ejaculation. When early testicular elevation occurs, and an extended excitement or plateau phase delays ejaculation, the contracted cremasteric musculature loses tone and allows spermatic cord relaxation. The cord lengthens and the testes descend toward the depths of the scrotum. When the delayed ejaculation does take place, the testes may be lower in the scrotal sac than they were during the extended excitement or plateau phase of the cycle. The aging male loses contractile tone of the cremasteric musculature much more rapidly than does the younger man and regains it at a much slower pace.

During excitement-phase levels of sexual tension, the testes of younger males frequently increase in size approximately 50 percent beyond the unstimulated baseline. This deep vasocongestive reaction continues through plateau and achieves its greatest severity as orgasm is experienced. Beyond the late fifties or early sixties, the human male rarely develops obvious vasocongestive increase in testicular size. In the present study, two males in the 51–60-year and one in the 61–70-year age group demonstrated occasional vasocongestive increase in testicular size.

Orgasmic Phase. There is no specific orgasmic-phase testicular reaction for the human male.

Resolution Phase. During the resolution phase, testicular descent in the aged male is so rapid that occasionally it has passed unobserved. Not infrequently, testicular descent starts during orgasmic experience. The testes may be returned to the scrotal depths before the full ejaculatory experience has been completed. Particularly is this true for men over 60 years of age.

Resolution-phase involution of the transitory increase in testicular size occurs almost as rapidly as the reaction of testicular descent in older men. It is so fleeting, in fact, that unless the testes are under direct observation, loss of the tension-inspired, localized, deep testicular vasocongestion may not be noted.

EJACULATION

The physiologic expulsion of seminal fluid by both the primary and the secondary organs of reproduction is the biologic expression of male orgasm. The act of ejaculation usually involves activity of the highest cortical levels, but also it may be purely reflex [110, 168, 207, 260, 285]. Following the established pattern of the younger male, the ejaculatory process of the aging male may develop in two stages: (1) the expulsion of seminal fluid from the accessory organs of reproduction into the urethral pars prostatica and (2) the progress of the seminal fluid through the urethra from the pars prostatica through the pars membranacea and the pars spongiosa to and through the urethral meatus.

Vocalization of subjective progression by young males has described the two-stage ejaculatory process as follows: The first stage, developed by accessory-organ contractions (prostate; questionably, the seminal vesicles; etc.), elicits the sensation of ejaculatory inevitability and the "feeling that the ejaculation is coming." In this situation the young male no longer voluntarily can constrain or control the ejaculatory process. There is a brief interval (2 to 3 seconds) when he feels the ejaculation coming, when he can no longer control it, but before the actual emission of seminal fluid occurs (see Chapter 14).

As the male ages, the entire ejaculatory process undergoes a reduction in physiologic efficiency. Ejaculation is altered not only in physiologic integrity but also in subjective progression.

In the older age group the first stage of ejaculation may be either foreshortened to the extent that all sensation of ejaculatory inevitability is lost or lengthened to such a degree as to stimulate spasmodically recurrent sensations of ejaculatory imperativeness. These marked variations in physiologic response will be explained in context.

If the aging male's penile erection has been maintained voluntarily during an extended excitement or plateau phase, the sensation of ejaculatory inevitability usually does not develop in an acute fashion. In this situation the male's ejaculatory process is one of sudden second-stage expulsion of seminal fluid through the urethral meatus without separate first-stage warning contractions of the accessory organs of reproduction. In other words, instead of a two-stage, well-differentiated ejaculatory process, the elderly male may have a single-stage expulsion of the seminal fluid, with the secondary organs of reproduction contracting simultaneously with the penile expulsive contractions rather than preceding them by 2 or 3 seconds. Thus, the physiologic process of ejaculation may be completed without a separate first-stage psychosexual warning of inevitability. Lowered ejaculatory pressure so frequently seen as the male ages also may contribute to reduction of sensual experience.

Rectal examinations made on one man in the 61–70 age group and one man over 70 years of age during ejaculation did not demonstrate clinically obvious contractions of the prostate during the ejaculatory process. Similar rectal examinations in younger men always have identified prostatic contractions which have onset during the first stage of the ejaculatory process and which continue into the second stage as well.

When the man over 60 retains first-stage contractions of the accessory organs, this stage may be markedly extended and his ejaculatory process altered significantly. The usual 2–3-second duration of the first stage may stretch into a period of 5 to 6 seconds during which the man has the feeling of ejaculatory imperativeness without the subsequent onset of second-stage penile contractions.

It is possible that the efficiency of ejaculatory function of the secondary organs of reproduction (prostate, seminal vesicles, etc.) may be impaired by the aging process. These organs, under stimulation, may develop a degree of spastic contraction rather than regularly recurrent expulsive contractions. A severe sensation of ejaculatory imperativeness develops, but the ultimate second-stage seminal-fluid emission may not occur. This form of secondary or acquired impotence usually is not a constant occurrence for the aging male. This ejaculatory inadequacy has been observed several times following long-continued penile erection and once in a 67-year-old male who was under the influence of alcohol.

In brief, the aging male with the sensation of ejaculatory imperativeness has the psychosexual demand to ejaculate, but there may be loss or inefficiency of neurophysiologic control of the process. It should be emphasized that this type of ejaculatory inadequacy usually is transitory in character.

During the second stage of the ejaculatory process (penile contractions) the young male can feel the actual emission of the seminal fluid and is aware not only of the force of the expulsive contractions but also of the localized sensation of fluid emission as the seminal plasma is expelled under pressure along the length of the penile urethra. The aging male, particularly if his erection has been long-maintained, may have the experience of seepage rather than of seminal-fluid expulsion. As the younger male is terminating the second stage of the ejaculatory process, a final drop or two of seminal fluid may seep from the urethral meatus, obviously without expulsive force. This same feeling of reduced or absent expulsive force may apply to the entire ejaculatory process of the man over 60 years of age. In this situation there are no identifiable first-stage contractions of the accessory organs of reproduction, and seminal fluid seeps from or escapes the urethral meatus without significant ejaculatory pressure. The expulsive penile contractions are reduced in intensity and duration. Such a male usually does not have the sensation of ejaculatory inevitability, nor is there the psychosexual satisfaction of expulsive penile contractions. For these elderly men the ejaculatory process is truly a one-stage rather than a two-stage affair.

As the male ages, the intensity and the duration of the localized vasocongestive processes are diminished. There is marked reduction in the frequency of the superficial sex flush and a slowing in the development of full penile erection. This deep vasocongestive phenomenon is delayed not only in rapidity but also in maturity of development. Testicular elevation and scrotal-wall vasocongestion are reduced in reactive quality, and congestive testicular size increase does not occur frequently beyond 55 years of age. The efficiency of ejaculatory physiology is reduced in reactive intensity in both first and second stages. The psychosexual pleasure of the ejaculatory process may be impaired. Finally, the two-stage ejaculatory pattern may be reduced to a single-stage, or other abnormal ejaculatory patterns may develop.

2 .   CLINICAL   CONSIDERATIONS

Our aging population increasingly demands functional and functioning roles not only within community structure but in private life. Vocalization of these demands has stimulated renewed consideration of the inevitable adjustments of the aging process. Not the least of these adjustments are those related to physiologic and psychologic capacity for sexual performance.

Investigative scrutiny rarely has been directed toward the functional and functioning variations of the sexuality of the human male during and beyond his climacteric years. Much in the aging male's life is affected by psychosocial adjustments occasioned by sexual involution. Yet little scientific data has been established as a baseline from which his psychosexual needs may be interpreted and solutions to related problems suggested. Although the Kinsey group recorded data on over 5,000 white males, only 126 histories were obtained from men past 60 years of age [142]. Both Finkle *et al.* [73] and Newman and Nicols [240] in independent studies interviewed just over one hundred males beyond the age of 55 years. Although Stokes [303] has assembled a massive amount of pertinent information, his statistics and conclusions have not been published to date [304].

In short, clinical material gathered to evaluate sexuality of males in the geriatric population has been totally inadequate. This statement should not be construed as reflecting adversely upon prior investigations, but rather should serve to emphasize the difficulties inherent in any attempt to evaluate the aging male's sexuality. In order to understand the rigid social resistance expressed toward any investigation of the aging male's sexuality, it may be helpful to recall that Victorian influence upon our society has decreed for years that the aging male possesses little or no socially acceptable sexuality.

When this culturally resistant facet of human sexual experience was approached, the anticipated inadequacy of available clinical material was encountered. In order to gather the 39 men (see

Part 1 of this chapter) who were able to cooperate unreservedly with the program, 245 men over the age of 50 years were interviewed. Of this total, 212 were sufficiently motivated to provide detailed sociosexual histories which dated from earliest recall to current state of sexual activity. Of the 212 aging men, 152 were members of marital units in which both partners were past 50 years of age and were willing to provide histories of sexual behavior. Full advantage was taken of the opportunity to cross-check the sociosexual histories of the 152 men by comparing them insofar as possible with those obtained from their wives. There was marked correlation in material on sexual capacity and performance, including specifics of current frequency, techniques, mounting success, and patterns of satisfactory response. When an aging male or female cooperated to provide a history, the partner was interviewed immediately to avoid suggested or jointly prepared answers. It is from a review of this material, together with data acquired from seven years of clinical therapy of male sexual inadequacy, that this chapter has been constituted.

The aging male's obvious reluctance to impart material with sexual orientation is evidenced by the fact that 245 men had to be interviewed in order to accumulate the 212 histories granted separately to both the male and female members of the interview team. The age distribution, by decades, of these 212 cooperative men and brief statistics relating to their formal education are presented in Table 16-2.

TABLE 16-2

Age and Education of 212 Geriatric Male Study Subjects

Age by Decade	No. Subjects	Education			
		Grade School	High School	College	Graduate School
51–60	89	2	17	43	27
61–70	71	0	8	51	12
71–80	37	1	6	26	4
81–90	15	1	4	9	1
Totals	212	4	35	129	44

The bias toward higher levels of formal education than would be expected for the general population is obvious. Only 4 men failed to enter high school, and only a total of 39 (18 percent) failed to enroll in college. Despite the exceptionally higher levels of formal education apparent in the male study-subject population, only 39 men had sufficient incentive and security of sexual performance to cooperate actively with the study-subject group (see Part 1 of this chapter).

There is no question of the fact that the human male's sexual responsiveness wanes as he ages. Particularly is this true if sexual responsiveness arbitrarily is defined in such general, clinical terms as (1) existing levels of sexual tension, (2) ability to establish coital connection, (3) ability to terminate coition with ejaculation, and (4) current history of masturbation and/or nocturnal emission. A major difference exists between the response patterns of the middle-aged male (41–60 years) and those of men past the 60-year landmark. This difference is reflected in the male over 60 years by loss of maintained levels of sexual tension and reduced reactive intensity during sexual expression. Not only does coital activity usually decrease but the incidence of masturbation and nocturnal emission also is slowed with advancing years [26, 142].

The aging male's sexual capacity and performance vary from individual to individual and from time to time in a particular individual. Obviously, capacity and performance are influenced directly by acute or chronic physical infirmity, or by the general physiologic involution of the total body. Possibly the greatest influence on geriatric sexual response may be inherent in the sociosexual environment within which a male lives during his sexually formative years. Kinsey et al. [142] suggested this possibility in 1948, and work with the present investigative population has tended to support their thesis.

The most important factor in the maintenance of effective sexuality for the aging male is consistency of active sexual expression. When the male is stimulated to high sexual output during his formative years and a similar tenor of activity is established for the 31–40-year age range, his middle-aged and involutional years usually are marked by constantly recurring physiologic evidence of maintained sexuality. Certainly it is true for the male geriatric sample that those men currently interested in relatively high levels of sexual expression report similar activity levels from their formative years. It does not appear to matter what manner of sexual expression has been employed, as long as high levels of activity were maintained.

The incidence of sexual inadequacy in the human male takes a sharp upturn after 50 years of age. As might be expected, secondary impotence increases markedly after this age and continues to increase with each additional decade. During the last seven years of experience in therapy for sexual inadequacy, 83 percent of impotent males have been past the age of 40 years at onset of treatment, and three of every four of these men were over 50 years of age. Of real interest is the fact that the male over 50 years old can be trained out of his secondarily acquired impotence in a high percentage of cases [39, 132, 133, 210, 216, 323].

Just as the secondarily impotent male over 50 years old can be reconstituted, so can the potent aging male's responsive ability, dormant for physical or social reasons, be restimulated, if the male wishes to return to active sexual practices and has a partner interested in sexual performance. If he is in adequate health, little is needed to support adequacy of sexual performance in a 70- or even 80-year-old male other than some physiologic outlet or psychologic reason for a reactivated sexual interest.

Briefly, if elevated levels of sexual activity are maintained from earlier years and neither acute nor chronic physical incapacity intervenes, aging males usually are able to continue some form of active sexual expression into the 70- and even 80-year age groups. Even if coital activity has been avoided for long periods of time, men in these age groups can be returned to effective sexual function if adequate stimulation is instituted and interested partners are available.

FACTORS IN MALE SEXUAL INVOLUTION

Under what physical conditions or psychic influences does the aging male progressively lose sexual responsiveness? Answers to these questions comprise information of extreme importance in both understanding and treating problems of geriatric sexuality. Although the sample is small and obviously is not representative, some suggestions worthy of consideration have emerged not only from intensive team interrogation of the aging male but in many instances from cross-interrogation of his spouse.

There are manifold physiologic and psychologic factors that contribute to involution of the aging male's sexual prowess. This becomes particularly apparent when interrogation is carried out in depth. Under detailed probing the individual basis for alteration in male responsive ability usually falls within one or more of six general categories: (1) monotony of a repetitious sexual relationship (usually translated into boredom with partner); (2) preoccupation with career or economic pursuits; (3) mental or physical fatigue; (4) overindulgence in food or drink; (5) physical and mental infirmities of either individual or his spouse; and (6) fear of performance associated with or resulting from any of the former categories. These six categories will be considered briefly but in specific terms.

MONOTONY IN SEXUAL RELATIONSHIP

Loss of coital interest engendered by monotony in a sexual relationship is probably the most constant factor in the loss of an aging male's interest in sexual performance with his partner. This monotony may be the end-result of a sexual relationship which did not develop beyond the stage of dutiful indulgence and/or physical need for tension release. It also may develop from a relationship in which the sexual component did not mature or keep pace with other facets of marital progression.

Since the factor of overfamiliarity with the partner does influence sexual responsiveness, it should be considered in some detail. For generations, the mores of a patriarchal society have not confined the male to one sexual partner. When this attitude exists in the male partner, it may provide a built-in reaction of sexual restlessness after many years of partner restriction in a monogamous state. The female partner may lose her stimulative effect as her every wish, interest, and expression become too well known in advance of sexual activity, especially if the subconscious male focus has anticipated multiple-partner sexual variation. Although the younger woman to whom such an aging male turns may not in fact be as effective a sexual partner from a purely physical point of view, the mere lack of familiarity with the new personality creates an illusion of variation so attractive to the sexual demands of many males. For the aging male, the natural concomitant of this unconscious drive for variation is the possible development of a need for a change of outlet in order to create sexual tension or stimulate sexual capacity.

The complaint of sexual boredom frequently originates in the fact that the female partner has lost sight of the necessity for working at the marital relationship with the same interests in stimulating and satisfying her male partner that she originally may have demonstrated at the outset of marriage. The female partner who incites boredom may have lost herself in the demands of children, in social activities, in an individual career, or in any combination of interests extraneous to the marital focus. By their own admission many of the women interviewed no longer showed either sexual interest in or sexual concern for their husbands. It is the attitude of being taken for granted that the male apparently rejects, at least at a subconscious level. The ego of the aging male is especially vulnerable to rejection, either real or illusional.

The female partner in her forties may age, from the point of view of physical appearance, more rapidly than the male partner. Her overemphasis on or poor handling of the menopausal years may impart an aura of being "unfemale," with the result that she ceases to have a sexually stimulative cathexis for the male (see Part 1 of Chapter 15). Any loss of interest in personal attractiveness joined with rapidity of menopausal disintegration of her basic physical appearance may contribute to the male's feeling either of rejection or of overfamiliarity.

MALE CONCERN WITH ECONOMIC PURSUIT

Most men in the 40–60-year age group are reaching the competitive heights of their occupations and are contending with the greatest personal or family needs. They are engrossed with striving for the preeminence in their particular interest deemed necessary to provide that level of family financial security demanded by their socioeconomic structure. The competitive male world can be and frequently is allowed to become a demanding, all-consuming structure. This major diversion of male interest outside the home progressively reduces time available to the marriage. The male who is immersed in such an economic pursuit may make little or no effort to vocalize his occupational concerns to his female partner. Maintaining communication at any level permits sexual interchange to remain a natural occurrence rather than to become the result of a major effort of physical and mental reorientation.

There are additional factors in the vagaries of the competitive male world that should be considered. When the male has had a bad day, when things have not gone well, usually there is less interest in sexual activity than when he has experienced a most successful day. While a small percentage of the men interviewed reported finding comfort and reconstitution of ego within the realm of sexual release, it is certainly true that preoccupation remains a major deterrent to male as well as female sexuality.

MENTAL OR PHYSICAL FATIGUE

Fatigue is an important element in the involution of male sexuality and exerts an ever-increasing influence during and beyond middle age. Mental rather than physical fatigue is the greater deterrent to male sexual responsiveness, although both are capable of major influence in lowering or aborting sexual tension.

If the middle-aged male has the type of employment that requires essentially a physical effort, he long since has conditioned his body to these physical demands and there usually is relatively little involution in sexual activity that results from such occupational efforts. The types of physical activity that have been reported directly to inhibit male sexual interest are those of unusual or excessive physical strain which are more often associated with recreation than with job demands. Rarely does the middle-aged male in our culture make any effort to maintain his physical being in good condition. Therefore, the type of activity associated with a weekend's recreation frequently is more exhausting than his routine job demands, particularly for the 50-plus age group. An aging male in poor physical condition, exposed to occasions of excessive physical activity, frequently complains of reduction in or complete loss of sexual responsiveness during the 24 to 48 hours immediately following such unaccustomed physical effort.

As stated, mental as opposed to physical fatigue is the greater deterrent to sexual tension during the male's middle-aged years. Reflected from the competitive male world, the loss of sexual interest paralleling "the bad day at the office" initially results from mental exhaustion which only later may be translated into a physical counterpart. As the male ages, anything with which he habitually is preoccupied necessitates the expenditure of significant mental energy and subsequently reduces his sexual responsiveness. Occupational, financial, personal, and family emergencies universally are reported by the male geriatric research population to repress severely any existent sexual interest not only during the immediacy of the emergency but also usually for significant lengths of time thereafter. This sensitivity of male sexuality to mental fatigue is one of the greatest differences between the responsiveness of the middle-aged and the younger male.

OVERINDULGENCE IN FOOD AND DRINK

The aging male's excessive consumption of either food or drink has a tendency to repress his sexual tensions as it also lowers his capacity to feel or achieve in other areas. Many males have reported diminution of intensity in sensual focus, sometimes to a degree of anesthesia, as a result of overeating. However, the repression of sexuality is transient in nature unless the individual's eating patterns are grossly excessive and on a constant basis.

The syndrome of overindulgence has particular application to alcohol. While under its influence, many a male of any age has failed for the first time to achieve or maintain an erection of the penis. Secondary impotence developing in the male in the late forties or early fifties has a higher incidence of direct association with excessive alcohol consumption than with any other single factor. When a man is traumatized by the inability to achieve or to maintain an erection while under the influence of alcohol, he frequently develops major concerns for sexual performance and rarely associates his initial disability with its direct cause.

Not only does high alcohol ingestion directly reduce sexual tension in the aging male but also it often places upon him the additional indirect burden of concern for performance. He usually faces this secondary psychologic problem, if it persists, either by partial withdrawal from or by total avoidance of marital sexual exposure. His frequent solution to his erective concerns is to seek a sexual source unfamiliar with his personal concerns of sexual inadequacy. If, coincidentally, he refrains from adding excessive alcohol ingestion to the occasion of the first coital opportunity with the new partner, his solution probably will work. Thus a new problem within the marriage may arise. He is impotent with his wife but has confidence in his sexual performance elsewhere.

The alternative to the alcohol-dependent male with an impotence pattern is the picture of the true alcoholic. As this male progressively deteriorates physically and mentally, his sexual tensions simply disappear. Since the chronic or even acute alcoholic is in reality a mentally infirm individual, sexual involution under these circumstances will be discussed under the next heading of physical and mental infirmities.

PHYSICAL AND MENTAL INFIRMITIES

Physical infirmities that can reduce or eliminate sexual capacity and performance obviously may develop at any age. However, onset incidence rises precipitously beyond the 40-year age mark and, of course, is particularly a factor after 60 years of age. Any physical disability, acute or chronic, may and usually does lower the sexual responsiveness of the involved male. If the physical distress is acute, pneumonia for example, lack of sexual tension usually is transient and is accepted without question by both husband and wife. If physical infirmity develops as a chronic or slowly progressive distress, emphysema for example, involution of sexual capacity is among the early debilitating effects of progressive reduction in physical efficiency. Longstanding chronic metabolic diseases such as diabetes are known for the high incidence of associated secondary impotence [104, 273].

Most forms of involutional psychopathology are associated with evidence of reduced sexual activity. There are, however, notable exceptions to the general rule of sexual regression for the male experiencing mental senility. The entire problem has had little definitive investigation.

In short, any acute or chronic distress, mental or physical in character, that reacts to impair the male's general physical condition or to reduce the efficiency of his body economy may be associated with lowered or absent sexual tensions.

Androgens and estrogen-androgen combinations are being used with increasing frequency in an effort to maintain the aging male in a positive protein balance, just as these steroids have been used in younger men with steroid imbalance [106]. There has been some evidence of reawakened sexual interest subsequent to effective steroid replacement in aging males. Clinical impression suggests that the obvious elevation of eroticism is not a direct effect of steroid replacement. Rather, it may be a secondary result of the obvious improvement in total body economy and of a renewed sense of well-being [201].

Beyond the 60-year age level, the physical infirmity of the female partner also is an ever-increasing factor. If the wife is physically infirm, the aging male is restricted in sexual opportunity. As previously mentioned, regularity of sexual expression is the key to sexual responsiveness for the aging male. With loss of sexual outlet, many aging males report rapid loss of sexual tension. It should be emphasized that this situation is not as acute for the aging husband with a physically infirm wife as it is for the aging wife with a physically infirm husband. In our culture, the aging man has much more opportunity for sexual outlet than does the aging woman.

FEAR OF FAILURE

There is no way to overemphasize the importance that the factor "fear of failure" plays in the aging male's withdrawal from sexual performance. Obviously, any of the categories discussed briefly above would and do create in the aging male a fear of ineffective sexual performance. Once impotent under any circumstance, many males withdraw voluntarily from any coital activity rather than face the ego-shattering experience of repeated episodes of sexual inadequacy. Not infrequently they vocalize, and eventually come to believe, extraneous excuses for sexual withdrawal rather than accepting the clinical fact of a normal involutionary process.

Expressions of anger or personal antipathy toward a partner frequently are used as an escape from a feared loss of ability to perform adequately. Innumerable instances of the middle-aged male's turning to a younger female partner for sexual stimulation provide everyday cases in point. This clinical picture has been interpreted widely as the male's subconscious attempt to reestablish sexual potency in his own eyes, and to support his ego by proving repeatedly his sexual prowess. The fallible element in this solution is obvious when attempts to meet the increased demands of the younger partner often have changed the aging male's passing concern for performance into an established physiologic sexual inadequacy.

When aging males express a lack of interest in sexual performance or seek sexual stimulation extraneous to the marriage, their wives are left without true insight into their husband's fear of performance and may feel personally rejected by his withdrawal from marital sexual activity. If insight is present, the wives still fear to push the reluctant aging male into the possibility of recurrent episodes of erective failure. In any event, attempts at sexual performance usually are reduced in intensity and frequency, and the real factor of sexual stagnation takes over the marriage. When the aging male is not stimulated over long periods of time, his responsiveness may be lost.

There is every reason to believe that maintained regularity of sexual expression coupled with adequate physical well-being and healthy mental orientation to the aging process will combine to provide a sexually stimulative climate within a marriage [131]. This climate will, in turn, improve sexual tension and provide a capacity for sexual performance that frequently may extend to and beyond the 80-year age level.

GENERALITIES IN
SEXUAL RESPONSE

17

SIMILARITIES IN
PHYSIOLOGIC RESPONSE

The anatomy and physiology of human response to sexual stimuli have been discussed in detail. In order to provide continuity, the presentation of material has been oriented primarily to anatomic structuring of reproductive viscera. While the technique of discussing physiologic capacity of the human male and female for sexual performance in separate presentation does improve continuity, it also tends to create a false impression. This technique tends to emphasize the differences in sexual response between the two sexes rather than the similarities.

Certainly there are reactions to sexual stimulation that are confined by normal anatomic variation to a single sex. There also are differences in established reactive patterns to sexual stimuli —for example, duration and intensity of response—that usually are sex-linked in character. However, parallels in reactive potential between the two sexes must be underlined. Similarities rather than differences of response have been emphasized by this investigation.

The similarities of male and female response to sexual stimuli are highlighted in this skeletal review of body systems and in the accompanying discussion of physiologic reaction presented within the frame of reference of the cycle of sexual response.

THE BREASTS

The first response of the female breast to sexual stimulation is erection of the nipple. However, the nipples frequently do not achieve a state of full erection simultaneously. This is particularly

273

true when there is an obvious difference between the breasts in volume or when the nipples in unstimulated positioning are either flat or in some degree of inversion. Usually the nipple of the smaller breast or that least inverted will be first to become fully erect. There has been no constancy of nipple erection established for the male breasts. A minimum of 30 percent of sexually responding men provided positive evidence of an erective reaction of the nipples in an earlier study [213]. A wider experience with the sexually responding human male of all ages has resulted in an upward revision of these figures. At present it is estimated that 50 percent to 60 percent of all sexually responding males demonstrate some degree of nipple tumescence, resulting in a partial, if not full, erective reaction.

As opposed to the excitement-phase timing for female nipple erection, that of the male, although frequently developing during the excitement phase, well may be delayed into the plateau phase before a full erective reaction is established. There is also a turgidity or tumescence of both the male and female nipple developing after full erection is achieved, which has not been mentioned previously. This increase in nipple diameter rather than in nipple length develops late in excitement or early in plateau phase for the human female, but not before plateau phase is well advanced for the human male. This turgidity of the nipples becomes particularly evident if either excitement or plateau phase of the sexual response cycle purposely is prolonged to an unusual degree. This tumescence of the fully erect nipple in response to long-maintained high levels of sexual tension is masked so frequently in the female by advanced tumescence of the areolae that no estimates of incidence can be presented with security. Since turgidity of the erect male nipples occurs, as in the female, only during long-maintained plateau levels of sexual tension, no satisfactory estimate of frequency of occurrence can be provided.

The increase in breast size and tumescence of the areolae, which are relatively constant clinical entities for the human female (particularly for the nullipara), are not demonstrated by the male breast.

As opposed to the relatively rapid resolution-phase involution of female nipple erection, male nipple erection, when established, may persist for many minutes or even hours after ejaculation before detumescence has been completed.

SEX FLUSH

Both male and female study subjects have demonstrated a superficial vasocongestive flush that develops in response to sexual stimulation. For the female the sex flush appears late in excitement or early in the plateau phase, but for the male the flush has been observed only late in the plateau phase of sexual response. The sex flush achieves its widest distribution in both sexes in the immediate preorgasmic sequence. It has been observed in approximately 75 percent of all female and 25 percent of all male study subjects during individual sexual response cycles. Environment contributes to the incidence of the sex flush. As might be expected, it is more apt to develop in a warm rather than a cool atmosphere. Despite acknowledged environmental influence, the appearance of the sex flush always identifies severe levels of sexual tension. For instance, a study subject may go through an entire orgasmic experience without manifesting a flush, yet, in an immediately subsequent sexual encounter during which more severe levels of sexual tension develop, a widespread sex flush may be quite evident.

In both sexes, when the flush occurs it originates over the epigastrium and spreads to the anterior chest wall. The neck, face, and forehead are involved progressively. In the female the flush frequently spreads over the lower abdomen, the thighs, the arms, and even the low back and the buttocks in the more extensive reactions. With the exception of the forearms and shoulders, the flush rarely is seen in the male other than in the primary sites of epigastrium, chest, neck, face, and forehead.

Involution of the sex flush during the resolution phase follows a well-established pattern of inverse order of occurrence, regardless of the sex of the individual involved. The flush first disappears from the epigastrium, chest, shoulders, extremities, and back, and finally from the neck, face, and forehead.

MYOTONIA

Since elevated muscle tension is second only to vasocongestion as physiologic evidence of eroticism, the numerous voluntary and involuntary reactions of skeletal muscles to progressive degrees of sexual tension are understandable. During the excitement phase of sexual response muscle tension is primarily voluntary in character. Some evidence of involuntary-muscle activity, such as expansion of the vagina in length and transcervical diameter (see Part 1 of Chapter 6) and partial testicular elevation (see Chapter 13) does occur. However, evidence of muscle tension increment is slow to develop in either sex when compared to the excitement-phase evidence of vasocongestive reactions.

In general, specific examples of both voluntary and involuntary muscle tension first develop in the plateau phase of the sexual response cycle. The musculature of the legs, arms, and abdomen, as well as the neck and face, contract or contort spasmodically as late plateau and orgasmic phases of sexual tension are experienced. One of the most prominent of the involuntary muscle tension reactions is that of carpopedal spasm, which develops late in excitement or early in plateau-phase response. Carpopedal spasm has been observed in both male and female study subjects in the supine position during coition. However, male and female subjects develop severe carpopedal spasm more frequently when the sexually stimulative techniques are manipulative rather than coital.

The physical activity of coition, with normal employment of the voluntary musculature of the trunk, pelvis, and extremities, precludes frequent development of uncoordinated striated-muscle spasm for the superior-positioned coital partner. However, many male partners have demonstrated involuntary contractile spasm of the gluteal musculature immediately prior to ejaculation when in a superior position. They also use voluntary gluteal contraction extensively during masturbation or when trying to avoid premature ejaculation. Preorgasmic contractions of the gluteal musculature also have been observed frequently in female study subjects. The female, even more than the male, uses voluntary gluteal contractions to elevate sexual tensions. This technique is prevalent in female superior or knee-chest coital positioning or during automanipulative episodes.

During the orgasmic phase both sexes regularly demonstrate involuntary spastic contraction of general muscle groups such as the rectus abdominis and sternocleidomastoid muscles and the facial musculature. The orgasmic spasticity of these muscle groupings extraneous to the pelvic musculature has been discussed in the literature for many years [8, 21, 76, 88, 141–143, 155, 228].

During resolution, muscle tension is lost rapidly for both sexes if orgasmic release is experienced. As a rule voluntary-muscle contraction is released more rapidly than involuntary-muscle spasm. If orgasm does not develop, muscle tension is lost with a rapidity that parallels the degree of sexual tension achieved. For instance, there is little muscle tension either voluntary or involuntary developed during the excitement phase, but significant tension of both types is accumulated during the plateau phase. No difference has been observed between the sexes in rapidity of muscle tension release.

HYPERVENTILATION

Hyperventilation is a constant late-plateau-phase reaction for both sexes, regardless of the type of sexual stimulation or activity [119, 239, 268, 308]. The physiologic intensity and duration of the reaction are indicative of the degree of sexual tension that has developed. Respiratory rates have been recorded at peaks as high as 40 per minute for both sexes.

When this physiologic evidence of sexual tension develops late in male plateau phase, it usually continues through orgasm and terminates during the refractory period of the resolution phase. The female has been observed occasionally to return to a second orgasmic experience before the primary state of hyperventilation has resolved. In other words, once the male, even a young man, ejaculates, he is refractory to reinstituted sexual stimulation until the hyperventilative reaction has subsided, but the female may move from first to second orgasm while still hyperventilating from the first orgasmic experience.

TACHYCARDIA

Both the sexually responding human male and female achieve equal degrees of tachycardia while experiencing the various levels of sexual stimulation. Recordable cardiac rates range from 100 to 175 beats per minute during plateau phases of sexual tension. During orgasmic experience cardiac rates have been recorded from 110 to 180+ beats per minute. Tachycardia developing during advanced degrees of sexual tension has been described frequently in the past [6, 22, 30, 93, 136, 151, 152, 258, 305].

BLOOD PRESSURE

Blood pressure elevation in response to sex tension increment has been roughly parallel for both sexes. The male systolic elevations have increased by 40–100 mm. Hg and diastolic elevations by 20–50 mm. Hg. The female elevations have ranged slightly lower, with increases of 30–80 mm. Hg systolic and 20–40 mm. Hg diastolic.

Specific investigative details of cardiorespiratory physiology in sexually responding human males and females will be reported in a separate monograph on sexual physiology.

PERSPIRATORY REACTION

Approximately one-third of both sexes develop an involuntary sweating reaction during the immediate postorgasmic segment of the resolution phase. This reaction may develop whether or not there has been marked physical activity during the sexual encounter and whether or not a sex flush has appeared. In male study subjects this perspiratory reaction usually is confined to the soles of the feet and the palms of the hands, although appearing occasionally on the trunk, head, and neck. In the female, perspiration is distributed more frequently over the back, thighs, and chest wall, with occasional spread to the forehead and upper lip. The appearance of a postorgasmic perspiratory reaction has been described previously in the literature [58, 144, 234].

PELVIC VISCERA

EXCITEMENT PHASE

The first responses of the pelvic viscera to the stimulation of sexual tensions are the erection of the penis and the production of vaginal lubrication. Both reactions primarily are vasocongestive in character. Penile erection usually is accomplished in 3–8 seconds in the sexually responding male under 40 years of age. Generally, younger males (21–30-year age group) achieve penile erection faster than males over the age of 30 years. The comparable female response to the onset of sexual stimulation is vaginal lubrication. This transudate-like material appears on the walls of the vagina in the woman under 40 years old in 10–30 seconds from onset of stimulation. Beyond the age of 40, the speed of attainment of full penile erection or vaginal lubrication generally is slowed, but under effective stimulation these reactions may continue into the 80-year age group.

Both erection and lubrication vary in reactive intensity if excitement-phase levels of sexual tension are prolonged. The male may lose full penile erection during long-maintained excitement-phase levels of sexual stimulation. Under similar circumstances the female may reduce or even stop production of vaginal lubrication. In both sexes these reactions are reversible, particularly when sex tension levels are stimulated and/or maintained by manipulative rather than coital techniques.

There is a second vasocongestive response to excitement-phase levels of sexual tension that is identical for both sexes. It is demonstrated in the male by thickening of the scrotal integument, with resultant flattening, constriction, and elevation of the scrotal sac. In the nulliparous female the major labia elevate and flatten against the perineum; in the multipara the labia separate from the midline and thicken from venous congestion. The minor labia simultaneously develop a two- to threefold vasocongestive increase in size. This results in an extension of the vaginal barrel except at the fourchette.

As the excitement phase progresses toward plateau, the male responds to increasing sexual tensions by moderate elevation of the testes. Simultaneously the testes undergo a deep vasocongestive size increase. Correspondingly, the female expands the inner two-thirds of the vagina, adding 2–3 cm. to the length of the vaginal barrel, and develops a two- to threefold increase in vaginal width at the transcervical diameter.

As is true for penile erection and vaginal lubrication, the physiologic responses of enlargement and elevation of the testes and expansion and extension of the vaginal barrel are reversible reactions during electively prolonged excitement-phase levels of sexual tension. Under such circumstances, the depths of a relaxing scrotal sac and the normal scrotal folding patterns may reappear. These reversible reactions of the testes and scrotum will occur even if penile erection is maintained. Providing the female has not been mounted, prolonged excitement-phase tension levels will result in reduction of the transcervical expansion and a concomitant decrease in the vaginal-barrel length. These reactions will occur even though advanced degrees of vasocongestion of the vaginal walls and the minor labia are maintained.

PLATEAU PHASE

If sexual tension is elevated to plateau-phase levels, the male reflects these higher tension levels with further vasocongestive response. Although the penis may have appeared fully erect, an obvious increase in penile circumference at the coronal ridge develops before ejaculation.

The female's physiologic response to plateau-phase levels of sexual tension also provides further evidence of deep pelvic vasocongestion with the development of an orgasmic platform in the outer third of the vagina. This vasocongestive reaction usually progresses to such an extent that the outer third of the vaginal barrel may be obtunded by at least a 50 percent reduction in diameter.

Further parallels of physiologic response of the pelvic viscera of both sexes to plateau-phase levels of sexual tension are evidenced by superficial vasocongestive reactions. The male response is an increased purple cast to the coronal area of the penis. This color change is inconsistent, even if orgasm is to ensue. Many men never demonstrate this color change, while others do so only occasionally.

As opposed to this inconsistent male indication of preorgasmic tension levels, the female develops such a marked discoloration of the minor labia that it has resulted in the designation of these tissues as "sex skin." The nulliparous minor labia turn a bright red, and those of the multipara may darken to a burgundy-wine color as plateau-phase sexual tensions move the woman toward orgasmic expression. Once the minor labia go through this sex tension color change, orgasm is inevitable if effective sexual stimulation is maintained without interruption. Many women achieve plateau-phase sexual tension levels but do not experience orgasmic-phase release. These women do not show sex-skin discoloration of the minor labia.

Further clinical responses which develop in parallel manner for both sexes during the plateau phase are mucoid secretory emissions. The male has a preejaculatory emission from the urethral meatus that has been identified with Cowper's glands. This material usually is restricted to two or three drops in volume. Occasionally, during long-maintained plateau-phase levels of sexual tension, almost 1 cc. of the material has been secreted and collected.

The female mucoid emission is provided by Bartholin's glands during the plateau phase. The function of this material has been identified previously with vaginal lubrication [30, 119, 228, 268, 305, 318, 319]. Bartholin's glands do not contribute to the lubrication of the vaginal barrel with the exception of the immediate area of the fourchette and then only subsequent to long-maintained plateau-phase levels of sexual stimulation. With prolonged coital activity, the production of vaginal lubrication is slowed, and continued fourchette protection is accomplished by the Bartholin's glands. In view of the plateau-phase timing and the mucoid consistency of both materials, biochemical definition possibly may identify them as basically similar substances.

ORGASMIC PHASE

The ejaculatory phenomenon is unique in the entire cycle of sexual response and is the essence of the male orgasmic experience. Orgasm in the male is initiated by contraction of the accessory organs of reproduction. The vas deferens, seminal vesicles, ejaculatory duct, and prostate contract with regularity to provide the seminal plasma necessary for the ejaculate total. Contraction of the secondary organs of reproduction is experienced by the male as an impression of ejaculatory inevitability. The sensation developing over a 2–3-second interval has been described as a feeling that the ejaculate is coming and can no longer be controlled. This anticipatory sensation is experienced during the collection of seminal plasma in the prostatic urethra.

When seminal plasma is compressed in the prostatic urethra, the internal sphincter of the bladder contracts to avoid retrograde flow into the urinary bladder. The external bladder sphincter and the membranous portion of the urethra relax, and the entire seminal pool is ejected along the penile urethra and through the urethral meatus under significant ejaculatory pressure. These ejaculatory contractions are established by the periurethral muscles as well as by the ischiocavernosus and bulbospongiosus muscles. The contractions have onset intervals of 0.8 second and continue at this spacing for two to three expulsive efforts. Thereafter the contractions are slowed not only in interval timing but in expulsive force. However, they have been observed to continue on an irregular basis and with little expulsive force for two to four more occasions. The male orgasmic phase is terminated with the cessation of expulsive contractions of the penile urethra (see Chapter 14).

Female orgasmic experience usually is expressed over a longer time sequence. It has onset with contractions of the uterine musculature and of the vaginal orgasmic platform. The uterine contractions develop in the fundus and move toward the lower uterine segment. Concomitant with the onset of uterine contractions is the development of contractions of the orgasmic platform in the outer third of the vagina. These two physiologic responses to overwhelming sexual tension provide the sensations of total pelvic contraction that the sexually oriented woman identifies with orgasmic expression.

Both uterine and vaginal-orgasmic-platform contractions have simultaneous onset in an overall pelvic response pattern. The orgasmic-platform contractions have onset at 0.8 second and continue for four to eight contractions before there is recordable slowing in the intercontractile interval. Two to four more contractions may be observed thereafter at irregular intervals and with diminishing contractile force. They correspond to the non-expulsive slowed contractions of the penile urethra at the termination of the male's ejaculatory response (see Part 1 of Chapter 6, and Chapter 8).

Further parallels of pelvic visceral response during orgasmic expression have been observed in both sexes. In the male the rectal sphincter contracts two to four times, with onset intervals of 0.8 second. These rectal contractions have been observed to parallel in time sequence the expulsive contractions of the penile urethra. For the female the external rectal sphincter also contracts two to four times at 0.8-second intervals, again parallel in time sequence with contractions of the vaginal orgasmic platform. An additional female response is the occasional (10–15 percent) development of irregular contractions of the external urethra which occur without recordable rhythmicity and recur two or three times at the most.

RESOLUTION PHASE

The male has a unique refractory period which develops as the last, irregular, nonexpulsive contractions of the penile urethra occur and is maintained until sexual tension in the male has been reduced to low excitement-phase levels of response. The female has no such refractory reaction. She generally maintains higher levels of stimulative susceptibility during the immediate postorgasmic period. She usually is capable of return to repeated orgasmic experience without postorgasmic loss of sexual tension below plateau-phase levels of response.

Due to the postorgasmic loss of stimulative susceptibility, the male pelvic viscera tend to lose superficial and deep vasocongestion more rapidly than do those of the human female. Loss of penile erection in the male occurs in two stages. The first stage evidences rapid loss of vasocongestion until the penis is perhaps one to one-and-a-half times enlarged. Second-stage penile involution is a slowed response, particularly if the excitement- and/or plateau-phase levels of the specific sexual response cycle have been prolonged markedly. If penile erection has been maintained for long intervals before ejaculation, the second-stage involution of penile vasocongestion is usually an extended process.

Loss of testicular vasocongestion and return of the testicles to the depth of the scrotum, together with loss of congestion of the scrotal integument and the reappearance of the scrotal folding pattern, occur parallel in time sequence with the rapid primary involution of penile erection.

The female loses the deep vasocongestion of the orgasmic platform and the superficial sex-skin discoloration of the minor labia more rapidly than any other postorgasmic involutionary process. Yet these resolution reactions do not occur as rapidly as primary involution of penile erection. Thereafter, loss of minor-labia vasocongestion and the return of the walls of the vaginal barrel to unstimulated width and length frequently are extended over many minutes, despite a most satisfying orgasmic experience. The loss of the vaginal barrel's deep vasocongestion is slowed when compared to the rapid loss of scrotal and testicular vasocongestion.

Parallels between the anatomic responses of the human male and female to effective sexual stimulation have been established. As an aid to comparison, Tables 17-1, -2, -3, and -4 have been provided and are self-explanatory. It is obvious from the descriptions and discussions that the primary physiologic reaction of either male or female to effective sexual stimulation is superficial and/or deep vasocongestion, and that the secondary reaction is one of increased myotonia, both voluntary and involuntary in character. Hyperventilation, tachycardia, muscle spasm, pelvic-musculature contraction, etc., are of secondary import when compared to widespread general-body and organ-system vasocongestion. The parallels in reaction to effective sexual stimulation emphasize the physiologic similarities in male and female responses rather than the differences. Aside from obvious anatomic variants, men and women are homogeneous in their physiologic responses to sexual stimuli.

Tables 17-1, -2, -3, and -4 follow on pages 286–293.

TABLE 17-1

Sexual Response Cycle of the Human Female—Extragenital Reactions

	I. Excitement Phase	II. Plateau Phase	III. Orgasmic Phase	IV. Resolution Phase
Breasts	Nipple erection; increased definition and extension of venous patterning; increase in breast size; tumescence of areolae	Turgidity of nipples; further increase in breast size; marked areolar engorgement	No observed changes	Rapid detumescence of areolae and involution of nipple erection; slower decrease in breast volume and return to normal venous patterning
Sex Flush	Appearance of maculopapular rash late in phase, first over epigastrium, spreading rapidly over breasts	Well-developed flush; may have widespread body distribution late in phase	Degree of flush parallels intensity of orgasmic experience (est. 75% incidence)	Rapid disappearance of flush in reverse order of its appearance
Myotonia	Voluntary-muscle tension; some evidence of involuntary activity (vaginal-wall expansion, tensing of abdominal and intercostal musculature)	Further increase in voluntary and involuntary tension; semispastic contractions of facial, abdominal, and intercostal musculature	Loss of voluntary control; involuntary contractions and spasm of muscle groups	Myotonia rarely carried more than 5 min. into phase but not lost as rapidly as many evidences of vasocongestion
Rectum	No observed reaction	Voluntary contraction of rectal sphincter as stimulative technique (inconsistent)	Involuntary contractions of rectal sphincter occurring simultaneously with contractions of orgasmic platform	No observed changes

Hyperventilation	No observed reaction	Appearance of reaction occurs late in phase	Respiratory rates as high as 40/min.; intensity and duration indicative of degree of sexual tension	Resolves early in phase
Tachycardia	Heart rate increases in direct parallel to rising tension regardless of technique of stimulation	Recorded rates average from 100 to 175 beats per min.	Recorded rates range from 110 to 180+ beats per min.; higher heart rates reflect more variation in orgasmic intensity for female than for male	Return to normal
Blood Pressure	Elevation occurs in direct parallel to rising tension regardless of technique of stimulation	Elevations in systolic pressure of 20–60 mm. Hg, diastolic 10–20 mm. Hg	Elevations in systolic pressure of 30–80 mm. Hg, diastolic 20–40 mm. Hg	Return to normal
Perspiratory Reaction	No observed reaction	No observed reaction	No observed reaction	Appearance of widespread film of perspiration, not related to degree of physical activity

TABLE 17-2

Sexual Response Cycle of the Human Female—Genital Reactions

	I. Excitement Phase	II. Plateau Phase	III. Orgasmic Phase	IV. Resolution Phase
Clitoris	Tumescent reaction of clitoral glans; vasocongestive increase in diameter of clitoral shaft; shaft elongation	Withdrawal of clitoral body (shaft and glans) from normal pudendal-overhang positioning and retraction against anterior body of symphysis	No observed changes	Return to normal positioning within 5 to 10 sec. after cessation of orgasmic-platform contractions; slower detumescence and loss of vasocongestion
Vagina	Appearance of vaginal lubrication within 10–30 sec. after initiation of any form of sexual stimulation; expansion and distention of vaginal barrel; vaginal-wall color alteration from normal purplish-red to darker, purplish hue of vasocongestion	Development of orgasmic platform at outer third of vagina; further increase in width and depth of vaginal barrel	Contractions of orgasmic platform starting at 0.8-sec. intervals and recurring 5–12 times; after first 3 to 6 contractions, intercontractile intervals lengthen and contractile intensity diminishes	Rapid detumescence of orgasmic platform; relaxation of vaginal walls; return to normal coloring (may take as long as 10–15 min.)
Uterus	Partial elevation of anteriorly placed uterus; development of corpus irritability	Full uterine elevation into false pelvis; cervical elevation produces tenting effect in midvaginal plane; increasing corpus irritability	Corpus contractions beginning in fundus, progressing through midzone, and expiring in lower uterine segment; contractile excursion parallels intensity of orgasmic experience; multipara, est. 50% size increase	Gaping of external cervical os which continues 20–30 min.; return of elevated uterus to unstimulated resting position in true pelvis; cervical descent into seminal basin

	Excitement Phase	Plateau Phase	Orgasmic Phase	Resolution Phase
Labia Majora	Nullipara: flattening, separation and anterolateral elevation of labia away from vaginal outlet Multipara: vasocongestive increase in diameter; slight lateral movement away from midline	Nullipara: labia may become severely engorged with venous blood during prolonged phase Multipara: further vasocongestive swelling depending upon degree of varicosity involvement	Nullipara: no observed reaction Multipara: no observed reaction	Nullipara: return to normal thickness and midline positioning Multipara: involution of labial vasocongestion
Labia Minora	Minor labial thickening and expansion extending vaginal barrel approximately 1 cm.	Occurrence of vivid color change ranging from bright red to deep wine color; this sex-skin reaction pathognomonic of impending orgasm	No observed reaction	Color change from deep or bright red to light pink within 10-15 sec.; loss of vasocongestive size increase
Bartholin's Glands	No observed changes	Secretion of drop or two of mucoid material aiding in lubrication of vaginal outlet during long-maintained coital connection	No observed changes	No observed changes

TABLE 17-3

Sexual Response Cycle of the Human Male—Extragenital Reactions

	I. Excitement Phase	II. Plateau Phase	III. Orgasmic Phase	IV. Resolution Phase
Breasts	Nipple erection (inconsistent and may be delayed until plateau phase)	Nipple erection and turgidity (inconsistent)	No observed changes	Involution of nipple erection (may be prolonged)
Sex Flush	No observed reaction	Appearance of maculopapular rash late in phase (inconsistent); originates over epigastrium and spreads to anterior chest wall, neck, face, forehead, and occasionally to shoulders and forearms	Well-developed flush; degree parallels intensity of orgasm (est. 25% incidence)	Rapid disappearance of flush in reverse order of its appearance
Myotonia	Voluntary-muscle tension; some evidence of involuntary activity (partial testicular elevation, tensing of abdominal and intercostal musculature)	Further increase in voluntary and involuntary tension; semispastic contractions of facial, abdominal, and intercostal musculature	Loss of voluntary control; involuntary contractions and spasm of muscle groups	Myotonia rarely carried more than 5 min. into phase but not lost as rapidly as many evidences of vasocongestion
Rectum	No observed reaction	Voluntary contraction of rectal sphincter as stimulative technique (inconsistent)	Involuntary contractions of rectal sphincter at 0.8-sec. intervals	No observed changes

Hyperventilation	No observed reaction	Appearance of reaction occurs late in phase	Respiratory rates as high as 40/min.; intensity and duration indicative of degree of sexual tension	Resolves during refractory period
Tachycardia	Heart rate increases in direct parallel to rising tension regardless of technique of stimulation	Recorded rates average from 100 to 175 beats per min.	Recorded rates range from 110 to 180 beats per min.	Return to normal
Blood Pressure	Elevation occurs in direct parallel to rising tension regardless of technique of stimulation	Elevations in systolic pressure of 20–80 mm. Hg, diastolic 10–40 mm. Hg	Elevations in systolic pressure of 40–100 mm. Hg, diastolic 20–50 mm. Hg	Return to normal
Perspiratory Reaction	No observed reaction	No observed reaction	Involuntary sweating reaction (inconsistent), usually confined to soles of feet and palms of hands	

TABLE 17-4

Sexual Response Cycle of the Human Male—Genital Reactions

	I. Excitement Phase	II. Plateau Phase	III. Orgasmic Phase	IV. Resolution Phase
Penis	Rapid occurrence of erection which may be partially lost and subsequently regained during a prolonged phase, or may be easily impaired by the introduction of asexual stimuli	Increase in penile circumference at coronal ridge; color change in coronal area (inconsistent)	Expulsive contractions of entire length of penile urethra; contractions start at 0.8-sec. intervals and after the first 3 or 4 are reduced in frequency and in expulsive force; minor contractions continue for several seconds	Detumescence occurs in 2 stages: (1) rapid loss of vasocongestion until penis is from 1 to 1½ times enlarged; (2) slower involution to normal state, usually extended process
Scrotum	Tensing and thickening of scrotal integument; flattening and elevation of scrotal sac	No specific reactions	No specific reactions	Rapid loss of congested, tense appearance of scrotum and early reappearance of integumental folding; sometimes delayed process
Testes	Partial elevation of both testes toward perineum accomplished by shortening of spermatic cords	Enlargement of testes to a 50% increase over their unstimulated noncongested state; elevation to a position of close apposition to perineum; full testicular elevation pathognomonic of impending ejaculation	No recorded reaction	Loss of vasocongestive increase in testicular size and full descent of testes into relaxed scrotum; may occur rapidly or slowly depending upon length of plateau phase

292

Secondary Organs	No observed changes	No observed changes	Contractions of secondary organs which develop sensation of ejaculatory inevitability and initiate ejaculatory process	No observed changes
Cowper's Glands	No observed changes	Have been suggested as source of preejaculatory emission of 2 or 3 drops of mucoid fluid; timing is essentially same as that of secretory activity of Bartholin's glands in female; active spermatozoa have been observed in this fluid	No observed changes	No observed changes

I 8

MYOTONIA IN
SEXUAL RESPONSE

Sex tension increment initiates two physiologic reactions that have total-body distribution in both the human male and the human female—vasocongestion and myotonia. These protean reactions have multiform response to sexual stimuli. Vasocongestion develops as both superficial and deep reactions of target organs and appears in either form in many other areas and organ systems of the body. Myotonia may be evidenced initially by muscular irritability and, in mature form, is expressed and then resolved either by regularly recurring contractions or by severe spasm of the involved musculature. Venous congestion develops and is released as an involuntary reaction, while myotonia may be both voluntary and involuntary in origin and resolution.

Muscle tensions develop in response to sexual stimuli irrespective of the sex of the responding individual. Generally, the degree of muscle tension increment is related to specific levels of sexual tension. As a rule, clinically obvious myotonia is confined to plateau-phase tension levels. Despite the generally superior muscular development of the average human male as compared to that of the female, sex-linked relationships have not been established for the general development of myotonic response to sexual stimuli. There are exceptions to this statement. These sex-linked specific myotonic responses are reflected in the target organs of the male and female pelves, and they have been described with proper identification throughout this text. The framework of the four phases of the sexual response .cycle will serve as a descriptive mechanism for the patterns of both contractile and spastic muscular reaction to sexual stimuli.

Both coition and automanipulation must be considered in order to present an adequate clinical picture of voluntary and involuntary myotonic response to sexual stimulation. If every possible sexual positioning and resultant specific muscular response were considered, the discussion would be almost endless and the level of confusion extreme. For the sake of descriptive clarity, coition will be presumed conducted with the female in the supine position, and automanipulation with the individual also in supine positioning.

Recording techniques have been confined to direct observation and to cinematography. Accurate physiologic estimates of total body movement and of contractile excursion of specific muscle groupings of sexually excited men or women have been impractical. Polygraphic recording of muscle action potentials [5, 91] has not been attempted in the laboratory.

EXCITEMENT PHASE

Total-body response to sex tension increment is characterized clinically by increasing restlessness, irritability, and rapidity of voluntary and involuntary movement. In responding to excitement-phase levels of sexual tension, physical movement primarily is voluntary in character. As tensions elevate during the excitement phase, the responding individual moves restlessly and with increasing rapidity. Under the stimulation of sex tension increment, there is a gradual transition from the slow stroking or gentle holding mannerisms of an initial stimulative approach to a more rapid, forceful, and purposeful muscular movement.

During excitement-phase progression, there is obvious clinical evidence of increased myotonia in the long muscles of both the legs and arms, some involuntary tensing of the abdominal musculature, and an increase in the involuntary contractile rate of the intercostal musculature, with elevation of the respiratory rates of the responding individuals.

PLATEAU PHASE

With plateau phase established, myotonic response is clinically obvious from forehead to toes of the responding individual. In reacting to elevated sexual tension levels, the individual frowns, scowls, or grimaces as facial muscles contract involuntarily in semispasm. There may be a spastic contraction of the musculature surrounding the mouth. Late in plateau phase the mouth may be opened involuntarily in a gasping reaction to hyperventilative demand. There is more of a tendency toward oral patency in coition than during masturbation. During automanipulation the jaws frequently are clenched spastically, restricting inhalation to the nasal passages. In turn, the nares flare in hyperventilative response.

The muscles of the neck (sternocleidomastoid group) contract involuntarily in a spastic pattern. As the result, the neck usually is held rigidly in midposition, but with orgasm imminent, minor degrees of opisthotonos may develop during an automanipulative sequence or for the supine coital partner. Usually there is an involuntary increase in the rate of contraction of the intercostals as respiratory rates rise to maximum.

The abdominal musculature (recti abdominis) tense involuntarily in semispastic contractile reaction during automanipulative episodes. The same abdominal musculature voluntarily contracts rhythmically to amplify the forceful pelvic thrusting movements of the coital partners at plateau-phase tension levels.

Clinical evidence of myotonia in the long muscles of the arms at plateau-phase levels of sexual tension depends on positioning in coition and on specific automanipulative techniques. The arms of the supine coital partner may express sexual demand. As demand elevates, the voluntary muscular response of holding the shoulders or back or upper arms of the superior-positioned partner changes to an involuntary clutching or grasping reaction. With orgasm imminent, clutching or grasping by the supine coital partner is a well-established response pattern.

The arms and hands of the superior-positioned coital partner usually are devoted to body support. Spastic contraction of the involved musculature is rare in performance of this voluntary muscular function. However, late in plateau, with the individual straining for orgasmic release, the biceps frequently have been observed in involuntary spasm.

In response to plateau tension levels during automanipulation, there is a voluntary increase in rate of hand and arm movement and an involuntary increase in pressure applied locally either on the mons area or along the shaft of the penis, as described in Part 2 of Chapters 5 and 12. The uninvolved hand may develop a clutching, clawlike, spastic contraction of the musculature (carpopedal spasm). This reaction also has been observed in one or both hands of the supine coital partner during the terminal stages of active coition immediately prior to orgasm. Carpopedal spasm of the hands develops during coition only if the individual has not established or has released any grasping or clutching of the superior-positioned partner. This reaction has been described in Chapters 3 and 11 in detail.

In addition to the recti abdominis, the muscles most actively concerned with pelvic thrusting during automanipulation and active coition are those of the buttocks, the gluteal musculature. Late in excitement or early in plateau the gluteus muscles may be used in a purely voluntary manner to increase subjective response to sexual stimulation. Many women contract the muscles of their buttocks during automanipulation or coition to elevate sexual tensions from excitement into plateau-phase levels of response. The gripping and constricting sensations that voluntary contractions of the gluteal muscles develop are a most effective method of elevating sexual tensions. Voluntary gluteal contractions also are used by males to elevate their tension levels. Some males have been observed to carry themselves in masturbation from late excitement through plateau and to ejaculation with regularly recurrent voluntary contractions of the gluteal musculature, without direct manipulation of the penile shaft [144].

The active thrusting and accommodation of male and female pelves in coition are controlled fundamentally by coordinated contractions of the gluteal, abdominal, deep pelvic, and thigh musculature. Male pelvic thrust and female pelvic accommodation initially are voluntary muscular attempts at sex tension increment. Late in plateau phase, immediately prior to orgasmic expression, the rapid forceful pelvic thrusting of either sex essentially is involuntary in character. For the male partner pelvic thrusting becomes so involuntary a reaction late in the plateau phase that the penis rarely is withdrawn more than halfway from the depth of the vaginal barrel before again being thrust deeply into the cul-de-sac. Female pelvic accommodation to this manner of penile thrusting follows a similar involuntary reaction pattern. The voluntary, full-excursion, pelvic thrusting of each sex so characteristic of the excitement-phase tension increment becomes involuntary, plateau-phase tension demand and results in spastic reduction in pelvic excursion. Marked reduction in the excursion of penile thrust or vaginal accommodation as coition is continued is pathognomonic of impending orgasmic tension release.

The thighs provide a major muscular contribution to effective sexual stimulation and in the process voluntarily and involuntarily evidence high levels of myotonia. With coital onset, the legs of the supine female partner are abducted from the midline and may be elevated. Each of these accommodative movements is initiated by voluntary contractions of individual muscle bundles.

As sexual tensions increase there is a female tendency toward adduction of the thighs toward the midline. There may be voluntary adduction of the legs primarily in an attempt to accommodate male penile thrusting and secondarily to elevate the sexual tensions. The voluntary leg placement of excitement-phase accommodation becomes almost a mechanism of involuntary demand as tensions elevate through plateau-phase response level. Some women additionally use leg elevation or spastic constriction of the thigh musculature for subjective tension elevation just as they employ the gluteal or abdominal muscular constriction.

For the superior-positioned male in coition, the legs, knees, arms, and elbows provide body support. A significant degree of voluntary muscle tension is necessary to support the male partner above the female. In addition the male occasionally may have to support some portion of the female's weight. Most of the male's structural musculature is in voluntary contraction to support both his, and, if necessary, portions of his partner's, weight during excitement-phase tension levels. However, during the plateau phase the supportive musculature frequently contracts in semispasm in an essentially involuntary manner.

When the male or female is in plateau-phase response to automanipulative activity, the long muscles of the legs usually are in spastic extension and are partially adducted toward the midline. The thighs of the male responding to automanipulative techniques usually evidence the same adductive tendency that has been described for the coitally active or manipulating female. The myotonic total of these voluntary contractions added to those of the gluteus and the rectus abdominis muscles aids many men and women to achieve high levels of sexual response. With orgasm imminent, the voluntary rhythmic contractions of the thigh muscles and the glutei turn into involuntary spasm that may remain through orgasm.

Carpopedal spasm also develops frequently as an involuntary hyperextension of the arch of the foot and as clawlike contractions of the toes. This reaction may occur for the supine partner in coition and for either male or female responding to automanipulative techniques. These reactions also have been discussed in Chapters 3 and 11.

ORGASMIC PHASE

In orgasm, there are severe levels of muscle tension evident throughout the body, and almost every clinical evidence of myotonia is involuntary in character. Since there may be significant subjective loss of conscious focus during an orgasmic experience, muscle strain from severe spastic contraction frequently is not identified at the moment it occurs. It is not at all unusual for a forcefully responding or multiorgasmic coital partner to be aware the next day of severe muscle aching of the arms, legs, back, or lower abdomen. Usually such an individual while involuntarily responding in orgasmic experience is unaware of the expended physical effort that has occasioned the muscle strain. In response to effective sexual stimulation individuals may accomplish feats of muscular coordination that would be unattainable in sexually unstimulated states.

The specifics of target-organ contraction in the male and female pelves during orgasm have been discussed separately in Chapters 6, 7, 8, 12, and 14 and mentioned briefly in other areas throughout the text. Therefore, no attempt will be made to describe these sex-linked examples of muscle tension out of context.

RESOLUTION PHASE

Rarely is clinically obvious muscle tension carried into the resolution phase more than five minutes past the termination of orgasmic experience, assuming that sexual stimulation does not recur. However, in other than the target organs, myotonia usually does not resolve as rapidly as superficial or deep vasocongestion.

19

STUDY-SUBJECT
SEXUALITY

Sexuality is a dimension and an expression of personality. Culturally, sexual aggression has been accepted as a mode of expression for the human male, an integral part of the "plumage" of his dominant role. Currently, eroticism has become so synonymous with maleness that it has progressed beyond acceptability to desirability. It is presumed that only physical defect or the depletions of the aging process will interfere with the male's innate erotic interests and his ability to respond to sexual stimuli. Two conceptual errors defeat these basic presumptions: First, any fear of performance, displeasing sensation, or sense of rejection affects male eroticism as much as it does the physiologic effectiveness of his response; and second, age does not necessarily deplete the male's physiologic ability for or psychologic interest in sexual performance (see Chapter 16).

The fact that these presumptions of unflagging male eroticism have endured in our society is demanding of some reverence. The perpetuation of these fallacies may be related to the encompassing principles that male sexual performance is necessary for procreation and that the human female inevitably supports any cultural demand that places or maintains the male in that role so necessary to her own ultimate biologic function.

The acceptance of eroticism in the human female is as variable as are the cultures in our society [38, 137, 230]. To date, a sexual role for the female in which she freely participates has not received total acceptance in Western culture, despite the currently nebulous status of the double standard. The incredible swing from yesterday's Victorian repression to today's orgasmic preoccupation has taken the human female but a few decades, and the shock of the transition has been imprinted deeply on our society.

What are the current norms of male and female sexuality? How do they differ from the picture created by the Kinsey sampling collected almost a quarter-century ago? How does the sexuality of male and female study subjects differ from that of our general population, and are there significant differences? These biologic and behavioral questions are of major moment. Unfortunately, they are questions for which there are no answers, because there are no established norms for male and female sexuality in our society. In fact, the strength with which individual cultures continue to exist in this country precludes the emergence of any dominant influence. The sociologic challenge of accumulating statistics reflecting current human sexual behavior comparable to the Kinsey-Pomeroy data [142, 144] has not been accepted. There are no unquestioned authorities or sources of reference for any area in the total of human sexuality with the exception of Gebhard's recent contribution pertaining to sex offenders [89].

Without established norms of human sexuality, there is no scale with which to measure or evaluate the sexuality of the male and female study-subject population. Therefore, material collected from this investigational experience is presented in discussion format. Comparisons may be drawn electively between the reported response patterns of this highly selected research population and the reader's personal experience or his concept of norms of human eroticism in today's society. From these prejudiced levels of comparison, there is no appeal at this time.

In eleven years, 382 female and 312 male study subjects have cooperated actively with the research program. The ages have ranged from 18 to 78 years for the women and 21 to 89 years for the men. The many aspects of selection inherent in this research population have been discussed in Chapter 2. However, the histories of two women and two men arbitrarily have been selected to exemplify sociosexual background frequently encountered among members of the study-subject population. Age, parity, and years of active cooperation will be reported as of July 1, 1965, or the date of the subject's separation from the program.

SUBJECT A

Subject A, a 26-year-old woman, has cooperated with the research program for three years. Family history records high-school matriculation for both parents and graduation for the father, who has worked as an electrician. The mother has worked irregularly outside the home as a domestic. There is no evidence of mutual regard or continuing marital participation between mother and father. One sibling, an older sister, married and left home at 16 years of age. Discussion of sexual material within the family was not permitted, and toilet and clothing privacy were demanded.

Subject A has a masturbatory history starting at age 10, with a frequency of two or three times a week maintained during puberty and voluntary reduction to about a once-a-month level in the midteens. She began dating at 13 years. Heavy petting began at once and continued until age 15, when she describes the first coital occasion. During the remainder of her high-school years coital exposure continued with regularity and with multiple partners. Contraception was practiced irregularly and varied from a vague attempt at "rhythm," to withdrawal, condoms, and an intravaginal sponge.

At 20 years she married a man eight years older, was separated at 22, and divorced in her twenty-third year. Contraception (condom or withdrawal) was practiced by male demand during the marriage. Coital connection developed four or five times a week as much at her instigation as from male interest. During the separation year she worked as a file clerk. There was severe limitation of heterosexual opportunity, and she returned to a masturbatory frequency of a minimum of twice a week as she reacted to the socially enforced isolation from sexual opportunity. One homosexual overture was made to her, which she rejected.

After the divorce, Subject A volunteered to join the study-subject population upon suggestion of her physician, and has cooperated actively with the investigation for three years. She entered the program for two stated reasons—first because of financial need and second for the socially secure opportunity for regular release from her sexual tensions. Aside from the usual anatomic and physiologic review, she has worked specifically with intravaginal-contraceptive evaluation and uterine-contractility pro-. grams.

Her history of sexual responsiveness at the start of her project cooperation is one of multiorgasmic return to automanipulation and an estimated 50 percent orgasmic return during intercourse. The estimated 50 percent orgasmic response to sexual-partner stimulus was Subject A's pattern in high school and during her marriage. Within the research design she repeatedly has been multiorgasmic during automanipulation, and her orgasmic response to artificial coition has averaged 85 to 90 percent of opportunities. There have been several occasions in the research environment during which she was multiorgasmic during artificial coition. She has reported consistent use of fantasy when exposed to this technique.

The actual orientation of this subject to demands of the program, aside from the initial interrogation, involved three sessions devoted to (1) environment and equipment accommodation, (2) a masturbatory sequence, and (3) artificial coital experience, during which she was quite at ease and fully orgasmic.

In brief, Subject A has no history of family direction or protection. Onset of dating and heavy petting was reported to have occurred immediately after the other sibling, an older sister, left home. During her teenage years she developed and maintained multiple sexual relationships. Although at most 50 percent orgasmic in coital opportunity, she preferred coition to masturbation despite the fact that during automanipulative episodes she usually was multiorgasmic. Automanipulative release was used in marriage only during episodes of partner separation or subsequent to repeated coital occasion without orgasmic tension release.

Since sexual activity had become a major factor in the girl's life, termination of the marriage placed her in a difficult sociosexual position. Although there were several sexual partners during the separation year and an increased masturbatory frequency, Subject A was well aware that she could not return to her high-school pattern of indiscriminate acceptance of multiple sexual partners without the strong possibility of jeopardizing socially her chances of a successful second marriage. Obviously, the research program has provided the opportunity for some regularity of tension release and, of extreme importance to Subject A, the social protection of anonymity.

She has explored the possibility of remarriage on two occasions during the past three years while cooperating with the program, but has felt that neither of the two opportunities would satisfy her primary interest in the potential security of a home and family. On both occasions she has voluntarily emphasized her relief that she did not have to evaluate the marital opportunities in a prejudiced state of sexual need.

Subject A has been selected from the 106 unmarried women who have cooperated with the research program. She represents the two reasons for joining the program most frequently vocalized by these subjects—financial demand and sexual tension. Her referral source (physician) ranks third in frequency as a method of female study-subject recruitment.

SUBJECT B

Subject B, a 31-year-old woman, is a college graduate and has been married for seven years. She has had one full-term pregnancy (three years ago) without complication. There was an abortion during the year prior to the successful conception. Contraception currently is practiced by means of oral medication.

Family background records the father as a college graduate and an insurance broker, and the mother as a high-school graduate. There are two older brothers, both of whom are college graduates, married, and living outside the area. Sexual material was discussed frankly within the family, although both toilet and clothing privacy were maintained. She describes a warm relationship between her mother and father.

Subject B has a masturbatory history with onset at 15 years, which continued intermittently through her teenage and college years without a firmly established frequency level. She always has been orgasmic, although rarely multiorgasmic, during these auto-manipulative episodes. Her first coital experience was at 17 years. She established a sexual connection during her last year in high school that developed into a once-a-week pattern of coital frequency. She was not orgasmic, did not practice contraception, and did not become pregnant.

While in college she had intercourse with three men, and there were numerous occasions of heavy petting with individual or mutual manipulation to orgasm. She was engaged to one of these men during her last year-and-a-half at college, again developing a coital frequency of approximately once a week and supplementing this with episodes of mutual manipulation to orgasm. In addition to a casual practice of "rhythm," condoms were used for contraception.

After Subject B graduated from college, she taught locally at grade-school level. There was occasional coital exposure and she maintained an irregular pattern of automanipulative activity. She married at 24 years of age, after nine months of courtship, during the last five months of which there was regularly recurring coition, mutual manipulation, and mouth-genital contact.

Her husband, a college graduate four years older than Subject B, is a junior executive in an industrial firm. His sexual history is one of multiple heterosexual experimentation (no homosexual experience) during both high-school and college years. His only active homosexual contact was in the service after college graduation and was accepted because he "just wanted to see what it was like." He was engaged to another girl before he met Subject B. During this engagement, coition, mutual manipulation, and mouth-genital contact were employed freely.

The Subject B family unit has cooperated with the program for five-and-a-half years and contributed, in addition to the general physiology program, to the pregnancy and sexual-response investigations. The unit also has offered cooperation in the cardiorespiratory investigative program. Their contributions both individually and as a family unit have been invaluable.

This family unit was selected as representative of the most frequent source of recruitment for the study-subject population. They volunteered their services after learning of work in progress by local report, expressing the hope that they could contribute in some manner to knowledge of human sexual response. Formal orientation was limited to two sessions subsequent to the individual history-taking. During the first session the unit was exposed to the atmosphere and equipment of the research laboratory. As demonstrated during the second session, their coital response under observation was excellent. The husband never has had erective or ejaculatory difficulty. Over the five-and-a-half years of program participation Subject B has been orgasmic in 85 percent of recorded coital opportunities, occasionally developing multiorgasmic response. She always has been orgasmic, although rarely multiorgasmic, during an automanipulative sequence.

Both members of the Subject B family unit were reviewed in depth early in 1965 after five years of cooperation with the program. Neither partner described any identifiable variation in individual or mutual sexual responsiveness in the privacy of their home as opposed to the research environment. There has been no erective or ejaculatory difficulty regardless of the environment, and Subject B's orgasmic return has not been altered by research equipment or personnel. This man and woman have stated categorically that they have found program cooperation of significant importance in their marriage. They have volunteered to continue in the research program so long as there is need for their contribution.

SUBJECT C

Subject C is an unmarried 27-year-old male who has completed a graduate-school education. He joined the program in 1956 and was separated in 1960. His family history records a graduate-school education for his father and college graduation for his mother. There are one older and one younger brother and an older sister, all of whom have at least a college degree. The family background, as expected, is one of multiple intellectual interests, varied avocations, and relative lack of parental restraint. Sexual material always has been discussed openly. There has been no more than token toilet or clothing restraint. Subject C describes more rapport with his mother than his father, who he states has little real interest in other than academic pursuit.

Subject C recalled onset of automanipulative activity in the thirteenth to fourteenth year, continuing with a frequency of once or twice a week during the teenage years. He has a history of several homosexual experiences. The first, as a 15-year-old boy, was a single occasion with an older male, and the second was with a college roommate during his nineteenth year. This episode lasted three months. Finally, during service connection, he took advantage of two single opportunities. He always has assumed the passive role.

His heterosexual experience had onset with routine dating at 15 years and with petting ultimately to ejaculation at 16 years. Intercourse, first experienced at 17, was continued irregularly during his last year in high school. During his college years (other than during his three-month homosexual episode), coital regularity was established at a frequency level of two to three times a month. There were several different partners.

Subject C was asked to join the research population while in postgraduate school and remained with the program for four years. Orientation required five sessions after initial interrogation. The first exposure to atmosphere and equipment in the laboratory was followed by two successful automanipulative episodes, the first slowly and the second rapidly accomplished. The first coital episode resulted in failure of erective performance, but the second was successful, although the erection was maintained with difficulty. Subsequently, Subject C had no trouble with automanipulative activity, but there were four failures with coital partners during program cooperation that entailed 22 opportunities over the four-year period. His functional coital difficulty was not in achieving but in maintaining penile erection under the stress of recording opportunity.

Although interrogated in depth, Subject C described no concern for these episodes of erective inadequacy, stating that he fully understood the problem and that erective inadequacy only developed when he was fatigued mentally from long periods of concentration. Physical expenditure appeared to have little effect on his sexual responsivity. He considered these occasions of erective inadequacy and his homosexual episodes all part of life experience and expressed interest in continuing cooperation with the research program on an indefinite basis. His homosexual exposures will not be discussed in this text (see Chapter 2). Subject C was separated from the program when the investigation of target-organ physiology of male sexual response was completed. There was no immediate evidence that a residual concern for performance might be a factor in future sexual activity.

As part of a long-range evaluation, Subject C was reinterviewed in August, 1965, approximately five years after separation from the program. He has married and is the father of two children. He describes his experience with the program as an important episode in his personal education and one which he feels has been most valuable to his marriage. He has had no erective difficulties during the marriage, nor does he describe any further homosexual interest or experience. Subject C has vocalized a desire to return to the program accompanied by his wife as a contributing family unit, when there is need for their services.

SUBJECT D

Subject D is a 34-year-old male, married 6 years and the father of two children. He dropped out of college at the end of his second year for both academic and financial reasons and has been working as a draftsman since college withdrawal. His family history is that of an only child. The father, a minister, died when the boy was 18 years of age. The mother, a high-school and secretarial-school graduate, has worked as a receptionist in a doctor's office since the father's death. There is a history of strict parental control during the teenage years and an excessive concern for social mores. There is no history of normal exposure to heterosexual activity until Subject D was a junior in high school. At that time dating not only was allowed but encouraged. However, curfews were strict and social control was dominant.

There is no automanipulative history until the first year in college, and this single experimental episode was followed for a full year by a severe guilt residual. In the second year of college there were several occasions of petting with one partner, once to involuntary ejaculation, with the experience followed for several months by residual guilt feelings. Overt sexual experience was essentially negligible until after Subject D left college and was working.

His first complete coital experience was with a prostitute one year after withdrawal from college. This episode developed into irregular prostitute exposure during the next four years. Social exposure was confined almost entirely to male acquaintances. The first history of steady dating developed at 25 years of age and lasted for six months. The connection was broken by the female partner because "she didn't think [he] was aggressive enough." During the next year there was a three-month casual social connection with a 20-year-old girl that resulted in several episodes of overt masturbatory behavior and two coital experiences. Shortly thereafter he met and seven months later married his wife. The premarital sexual history is one of heavy petting after three months of courtship and four coital episodes before marriage. Contraception was not used and pregnancy did not occur. Since marriage, conception has been controlled, first by diaphragm and later with oral medication.

Subject D's wife is a 33-year-old high-school and business-school graduate. She has a history of occasional automanipulative experiences starting late in her teens and continuing until marriage. During the first six years of her twenties she had coition with three men. Two men provided only single opportunities, but the relationship with the third man was maintained for three months. Premaritally she was orgasmic with automanipulation but not during coition. She met Subject D while working as a secretary in the same office with him. They have two children, 4 and 2½ years of age. There is a coital-frequency pattern of once or twice a week.

For orientation, five episodes were necessary after team interrogation. The first exposure was to background and equipment; during the second, coition was attempted without ejaculatory success. The third episode developed as successful coition for Subject D, but his wife was not orgasmic. During the fourth session both husband and wife were successful in individual automanipulative episodes, and in the fifth episode no difficulty was encountered by either partner in response to coital or manipulative stimuli. This pattern of freedom from restraint has been the rule in subsequent program exposure, much to the surprise of both husband and wife. The family unit has cooperated with long-range response-evaluation and male-physiology programs.

Subject D and his wife represent yet another important aspect in the development of the study-subject population. This family unit has, as have many others, volunteered their services, hoping to acquire knowledge to enhance the sexual component of their marriage in return for their cooperation with the program. Subject D was concerned with his own overcontrolled background, lack of sexual experience, and a considered degree of sexual repression. When his wife first joined the program, her orgasmic incidence was reported as rare with coital exposure but consistent with manipulation. Subsequent to working in the program her orgasmic achievement during coition has risen to the 80 percent frequency level. She has not been multiorgasmic either during coition or automanipulation.

Subject D's wife has stated repeatedly that subsequent to program participation her husband has been infinitely more effective both in stimulating and in satisfying her sexual tensions. He in turn finds her sexually responsive without reservation. Her freedom and security of response are particularly pleasing to him. Together they maintain that they have gotten a great deal more out of cooperating with the program than they have contributed, and they wish to continue on a long-term basis.

PROGRAM INFLUENCE ON STUDY-SUBJECT SEXUALITY

There are so many variables of sexual response that no possibility exists for establishing norms of sexual performance for the study-subject population. This position must be taken despite the degree of selection inherent in the requirement that there be a positive history of masturbatory and coital orgasmic experience before any study subject is accepted in the program. Many family units and individual men and women have joined the program for reasons which reflect basic concern with personal levels of sexuality. For example, one of Subject A's major reasons for participating in the program for the last three years has been the opportunity provided for anonymous relief of sexual tension. On the opposite side of the coin, Subject D and his wife joined the program for the express purpose of elevating the levels of their sexual responsiveness and resolving personal concerns created by inexperience and inadequacy of sexual performance.

From the onset of the investigation, major importance has been attached to the theoretical possibility that participation in the program might exert an adverse influence in the future upon the individual's own innate personal eroticism. There always has been the possibility that sexual activity conducted in the artificial atmosphere of the laboratory might create a responsive concern or fear of performance that could carry over into life experience. It also has been suggested that successful participation in the program might elevate sexual tensions to a degree incompatible with the sociosexual background of the individual involved. After eleven years there was no information available to suggest that active cooperation with the investigation has done other than maintain or improve the effectiveness of individual sexual expression.

Family units have been interrogated each year during their cooperation with the program, and after separation from the program are evaluated at five-year intervals. Since the investigation has been in existence eleven years, significant study-subject follow-up now is available. There has been no evidence of inadequacy in any phase of sexual performance developing among members of the study-subject population subsequent to exposure to the research environment.

Certainly, during episodes of active cooperation with the program there have been many observations of orgasmic failure or performance inadequacy. When failure statistics of coital performance were evaluated, it was interesting to note that the occasions of male sexual inadequacy far outnumbered those instances recorded for the female. When individual study subjects were unable to achieve ejaculatory or orgasmic levels of sexual tension in the research environment, the failure of performance was in the male on 65.1 percent of the occasions. These statistics reflect the tensions that accrue to the male and not to the female under the extremes of long-established cultural demand for effective sexual performance. Subject C, for example, failed to maintain erective adequacy in 4 of 22 coital exposures, although there is no history of automanipulative failure.

Beach [9, 13] has pointed to similar performance inadequacies in male cats and dogs in a laboratory environment.

Since episodes of sexual failure may create concern for future performance, some discussion of areas of failure is in order. Over the past eleven years, in automanipulative, coital, and artificial-coital activity, over 7,500 complete cycles of sexual response have been developed by female study subjects and more than 2,500 ejaculatory experiences have been recorded. During this period there were 338 failures of coital and automanipulative sexual performance in the research laboratory. Of the total of 338, 17 failures were with automanipulation (7 male and 10 female) and 321 recorded failures of orgasmic return were with coital activity (213 male and 108 female).

As expected, the highest concentration of failures of performance was encountered during the orientation program. However, neither the failures nor the successes of performance occurring during this period have been included in the reported statistics. In addition, there were many investigative episodes during which orgasmic return from the study-subject participants was not required. Such situations also have not been included in the performance statistics.

Primarily, male failure in coital exposure has been concentrated in the areas of erective inadequacy. Failures have developed either in attaining or maintaining penile erection to a degree sufficient for mounting effectiveness. As a secondary source of coital failure, premature ejaculation occasionally has been a problem with new members of the male population. Fortunately, this sexual inadequacy has been readily reversible. Premature ejaculation has not been of continuing concern, once adequate technical and clinical suggestions have been made, accepted, and practiced.

Female study subjects' orgasmic inadequacy also has been primarily coitally oriented. As stated in Chapter 9, female orgasmic experience usually is developed more easily and is physiologically more intense (although subjectively not necessarily as satisfying), when induced by automanipulation as opposed to coition.

The psychologic and sociologic variables that tend to affect female more than male sexual performance have been discussed in Chapter 9. In addition, the influence of the hormonal cycle and the depressant effect of excessive physical or mental fatigue always must be considered in relation to the effectiveness of sexual performance. However, for the female with manipulative and coital orgasmic experience fear of overt demand for orgasmic performance is a minor factor. The woman's fear of sexual performance which has been developing so rapidly in our culture in recent years certainly does not apply to the sexually experienced female. This is the major difference in reaction to the pressures of performance between male and female study subjects.

When female orgasmic or male ejaculatory failures develop in the laboratory, the situation is discussed immediately. Once the individual has been reassured, suggestions are made for improvement of future performance. As previously stated, there has been no evidence that occasional inadequacy of sexual performance under the pressures of the research environment has been transferred into a pattern of sexual inadequacy in private response. Exactly the opposite effect has developed. Many family units, following physician referral, have joined the research population in an effort to improve the effectiveness of their individual and mutual sexual performances. That these units have elevated their levels of sexuality is evidenced by their interest in and desire to continue as active research participants. There has been no evidence among the individuals and the family units separated from the investigation that the suggestions and techniques made available during program orientation have become other than an integral part of their private patterns of response.

Any assumption that definitive sexual stimulation accrues directly from exposure to research personnel or environment seems contradicted by the fact that overt exhibitionism has not been a factor in the laboratory. In fact, modesty, social control, and even an excessive regard for social mores has been the general response pattern.

Sexuality has many facets and many levels within the individual man or woman. The mercurial tendency to shift rapidly from peak to valley has been exemplified by female study subjects, while levels of sexual expression that remain essentially constant are observed most frequently in male study subjects. Through the years of research exposure, the one factor in sexuality that consistently has been present among members of the study-subject population has been a basic interest in and desire for effectiveness of sexual performance. This one factor may represent the major area of difference between the research study subjects and the general population.

REFERENCES

1. Adams, C. R. *Some Factors Relating to Sexual Rseponsiveness of Certain College Wives.* University Park, Pa.: Pennsylvania State University, 1953.
2. Allen, W. M., and Masters, W. H. Traumatic laceration of uterine support. *Amer. J. Obstet. Gynec.* 70:500–513, 1955.
3. Aristotle Problems. Bk. 2 and Bk. 4, in the *Works of Aristotle,* Vol. IV: *Historia Animalium* (D. W. Thompson, Trans.). Oxford, England: Clarendon Press, 1910.
4. Baldwin, J. F. The formation of an artificial vagina by intestinal transplantation. *Ann. Surg.* 40:398–403, 1904.
5. Balshan, I. D. Muscle tension and personality in women. *Arch. Gen. Psychiat.* 7:436–448, 1962.
6. Bartlett, R. G., Jr. Physiologic responses during coitus. *J. Appl. Physiol.* 9:469–472, 1956.
7. Baruch, D. W., and Miller, H. *Sex in Marriage.* New York: Harper, 1962.
8. Bauer, B. A. *Women and Love* (E. S. Jerday and E. C. Paul, Trans.). New York: Liveright, 1927.
9. Beach, F. A. Central nervous mechanisms involved in the reproductive behavior of vertebrates. *Psychol. Bull.* 39:200–226, 1942.
10. Beach, F. A. A review of physiological and psychological studies of sexual behavior in mammals. *Physiol. Rev.* 27:204–307, 1947.
11. Beach, F. A. *Sexual Behavior in Animals and Man* (Harvey Lectures, 1947–48, Series 43). Springfield, Ill.: Charles C Thomas, 1950.
12. Beach, F. A. Characteristics of Masculine Sex Drive. In *Nebraska Symposium on Motivation.* Lincoln, Neb.: University of Nebraska Press, 1956. Pp. 1–32.
13. Beach, F. A. Neural and Chemical Regulation of Behavior. In *Biological and Biochemical Bases of Behavior* (H. F. Harlow and C. N. Woolsey, Eds.). Madison, Wis.: University of Wisconsin Press, 1958.
14. Beauvoir, S. *The Second Sex.* (H. M. Parshley, Trans.). New York: Knopf, 1952.

15. Becker, H., and Hill, R. *Family, Marriage, and Parenthood.* Boston: Heath, 1948.

16. Benedek, T. Sexual Functions in Women and Their Disturbance. In *American Handbook of Psychiatry* (S. Arieti, Ed.). New York: Basic Books, 1959.

17. Benedek, T. Panel report: Frigidity in women. *J. Amer. Psychoanal. Ass.* 9:571–584, 1961.

18. Bergler, E. Frigidity in the female: Misconceptions and facts. *Marriage Hygiene* 1:16–21, 1947.

19. Best, C. H., and Taylor, N. B. *The Physiologic Basis of Medical Practice.* Baltimore: Williams & Wilkins, 1965.

20. Bliss, E. L. (Ed.). *Roots of Behavior.* New York: Paul B. Hoeber, Inc., 1962.

21. Bloch, I. *The Sexual Life of Our Time in Its Relations to Modern Civilization.* London: Rebman, 1908.

22. Boas, E. P., and Goldschmidt, E. F. *The Heart Rate.* Springfield, Ill.: Charles C Thomas, 1932.

23. Bonaparte, M. Notes sur l'excision clitoridectomy. *Rev. Franc. Psychanal.* 12:213–231, 1948.

24. Bonaparte, M. *Female Sexuality.* New York: International Universities Press, 1953.

25. Bors, E., and Comarr, E. Neurological disturbances of sexual function. *Urol. Survey* 10:191–222, 1960.

26. Botwinick, J. Drives, Expectancies, and Emotions. In *Handbook of Aging and the Individual, Psychological and Biological Aspects* (J. E. Birren, Ed.). Chicago: University of Chicago Press, 1959.

27. Brecher, E., and Brecher, R. Having a baby your own way: A guide to a healthy pregnancy. *Redbook* 122:69–80, Feb., 1964.

28. Brierley, M. Some problems of integration in women. *Int. J. Psychoanal.* 13:433–448, 1932.

29. Brown, D. Female orgasm and sexual inadequacy (a survey of the literature). Presented at a Conference of American Association of Marriage Counselors, Chicago, 1964.

30. Brown, F. R., and Kempton, T. *Sex Questions and Answers.* New York: McGraw-Hill, 1950.

31. Bryan, A. L., and Counseller, V. S. One hundred cases of congenital absence of the vagina. *Surg. Gynec. Obstet.* 88:79–86, 1949.

32. Bühler, C. *The First Year of Life.* New York: John Day, 1930.

33. Bunch, M. E. The concept of motivation. *J. Gen. Psychol.* 58:189–205, 1958.

34. Buxton, C. L., *et al.*   Bacteriology of the cervix in human sterility. *Fertil. Steril.* 5:493–514, 1954.
35. Buxton, C. L., and Southam, A. L.   *Human Infertility.* New York: Paul B. Hoeber, Inc., 1958.
36. Calderone, M. S.   *Release from Sexual Tensions.* New York: Random House, 1960.
37. Calderone, M. S. (Ed.)   *Manual of Contraceptive Practice.* Baltimore: Williams & Wilkins, 1964.
38. Calverton, V. F., and Schmalhausen, S. D. (Eds.)   *Sex in Civilization.* New York: Macaulay, 1929.
39. Carmichael, H. T., and Noonan, W. J.   The effects of testosterone propionate in impotence. *Amer. J. Psychiat.* 97:919–943, 1941.
40. Caufeynon, Dr. (pseudonym of Fauconney, J.)   *Orgasme: Sens Genital Jadis et Aujourd'hui.* Paris: Charles Offenstadt, 1903.
41. Chadwick, M.   *The Psychological Effects of Menstruation.* New York: Nervous and Mental Disease Pub. Co., 1932.
42. Chideckel, M.   *Female Sex Perversion.* New York: Eugenics Pub. Co., 1935.
43. Clark, L.   Female sex sensation. *Sexology* 25:208–212, 1958.
44. Clark, L.   Sexual Adjustment in Marriage. In *Encyclopedia of Sexual Behavior* (A. Ellis and A. Abarbanel, Eds.). New York: Hawthorn Books, 1961.
45. Counseller, V. S.   Congenital absence of the vagina. *J.A.M.A.* 136:861–866, 1948.
46. Covarrubias, M.   *Island of Bali.* New York: Knopf, 1947.
47. Cruickshank, R., and Sharman, H.   The biology of the vagina of the human subject: Part I. Glycogen in the vaginal epithelium and its relation to ovarian activity. *J. Obstet. Gynaec. Brit. Emp.* 41:190–207, 1934.
48. Dahl, W.   Die Innervation der Weiblichen Genitalien. *Z. Geburtsh. Gynaek.* 78:539–601, 1915.
49. Danesino, V., and Martella, E.   Modern concepts of functioning of cavernous bodies of vagina and clitoris. *Arch. Ostet. Ginec.* 60:150–167, 1955.
50. Davis, K.   *Factors in the Sex Life of 2200 Women.* New York: Harper, 1929.
51. Davis, M.   *The Sexual Responsibility of Woman.* New York: Dial Press, 1956.
52. Davis, M. E.   Estrogen and the Aging Process. In *Year Book of Obstetrics and Gynecology, 1964–65.* Chicago: Year Book Medical Publishers, 1965.

53. Deutsch, H. *The Psychology of Women*, Vols. 1 & 2. New York: Grune & Stratton, 1945.

54. Dickinson, R. L. *Human Sex Anatomy*. Baltimore: Williams & Wilkins, 1933.

55. Dickinson, R. L. Medical Reflections upon Some Life Histories. In *The Life of the Unmarried Adult* (I. S. Wile, Ed.). New York: Vanguard Press, 1940. Pp. 201–202.

56. Dickinson, R. L. *Atlas of Human Sex Anatomy* (2nd ed.). Baltimore: Williams & Wilkins, 1949.

57. Dickinson, R. L., and Beam, L. *A Thousand Marriages*. Baltimore: Williams & Wilkins, 1931.

58. Dickinson, R. L., and Beam, L. *The Single Woman*. Baltimore: Williams & Wilkins, 1934.

59. Dickinson, R. L., and Pierson, H. H. The average sex life of American women. *J.A.M.A.* 85:1113–1117, 1925.

60. Doyle, J. B. Exploratory culdotomy for observation of tubo-ovarian physiology at ovulation time. *Fertil. Steril.* 2:475–484, 1951.

61. Doyle, J. B. Ovulation and effects of selective uterotubal denervation: Direct observation by culdotomy. *Fertil. Steril.* 5:105–130, 1954.

62. Duvall, E. M. *Family Development*. Philadelphia: Lippincott, 1957.

63. Eissler, K. On certain problems of female sexual development. *Psychoanal. Quart.* 8:191–210, 1939.

64. Ellis, A. Is the Vaginal Orgasm a Myth? In *Sex, Society and the Individual* (A. P. Pillay and E. Ellis, Eds.). Bombay, India: International Journal of Sexology, 1953.

65. Ellis, A. *American Sexual Tragedy*. New York: Twayne Publishers, 1954.

66. Ellis, H. *Man and Woman*. Boston: Houghton Mifflin, 1929.

67. Ellis, H. *Studies in the Psychology of Sex*. New York: Random House, 1936.

68. Ellis, H. *Sex and Marriage*. New York: Random House, 1952.

69. Ellis, H. *Psychology of Sex* (2nd ed.). New York: Emerson Books, 1954.

70. Epstein, L. M. *Sex Laws and Customs in Judaism*. New York: Bloch Pub. Co., 1948.

71. Erikson, E. H. *Childhood and Society*. New York: W. W. Norton, 1950.

72. Falls, F. H. Simple method of making an artificial vagina. *Amer. J. Obstet. Gynec.* 40:906–917, 1940.

73. Finkle, A. L., *et al.* Sexual potence in aging males. *J.A.M.A.* 170: 113–115, 1959.

74. Fletcher, R. *Instinct in Man in the Light of Recent Work in Comparative Psychology.* London: Allen & Unwin, 1957.

75. Ford, C. S. *A Comparative Study of Human Reproduction.* New Haven, Conn.: Yale University Press, 1945.

76. Ford, C. S., and Beach, F. A. *Patterns of Sexual Behavior.* New York: Paul B. Hoeber, Inc., 1951.

77. Frank, R. T. The formation of an artificial vagina without operation. *Amer. J. Obstet. Gynec.* 35:1053–1055, 1938.

78. Frank, R. T., and Geist, S. H. The formation of an artificial vagina by a new plastic technique. *Amer. J. Obstet. Gynec.* 14:712–718, 1927.

79. Frazer, J. G. *The Golden Bough.* London: Macmillan, 1929.

80. Freeman, H., and Pincus, G. Endocrine Agents Other Than the Adrenal Forms. In *Clinical Principles and Drugs in Aging* (J. T. Freeman, Ed.). Springfield, Ill.: Charles C Thomas, 1963.

81. Frenkel-Brunswik, E. Studies in biographical psychology. *Character & Personality* 5:1–34, 1936.

82. Freud, S. Instincts and Their Vicissitudes. In *Collected Papers* (1921), Vol. IV. London: Hogarth Press, 1946. Pp. 60–83.

83. Freud, S. *New Introductory Lectures on Psychoanalysis* (W. J. H. Sprott, Trans.). New York: W. W. Norton, 1933.

84. Freud, S. *A General Introduction to Psychoanalysis* (J. Riviere, Trans.). New York: Perma Giants, 1935.

85. Freud, S. Beyond the Pleasure Principle. In *Complete Psychological Works of Sigmund Freud* (Standard ed.) (J. Strachey, Trans.), Vol. 18. London: Hogarth Press, 1955.

86. Froimovich, J., *et al.* Estrogenic Response of the Vagina in Geriatric Patients. In *Medical and Clinical Aspects of Aging* (H. Blumenthal, Ed.). New York: Columbia University Press, 1962.

87. Fulton, J. F. (Ed.) *Howell's Textbook of Physiology.* Philadelphia: W. B. Saunders, 1955.

88. Gardner, W. U. Reproduction in the Female. In *Textbook of Physiology* (J. F. Fulton, Ed.). Philadelphia: W. B. Saunders, 1950.

89. Gebhard, P. H., *et al.* *Sex Offenders.* New York: Harper & Row and Paul B. Hoeber, Inc., 1965.

90. Golden, J. S. Management of sexual problems by the physician. *Obstet. Gynec.* 23:471–474, 1964.

91. Goldstein, I. B., *et al.* Study in psychophysiology of muscle tension: I. Response specificity. *Arch. Gen. Psychiat.* 11:322–330, 1964.

92. Grad, H.   The technique of formation of an artificial vagina. *Surg. Gynec. Obstet.* 54:200–206, 1932.

93. Grafenberg, E.   The role of urethra in female orgasm. *Int. J. Sexol.* 3:145–148, 1950.

94. Gray, H.   *Anatomy of the Human Body* (W. H. Lewis, Ed.) (23rd ed.). Philadelphia: Lea & Febiger, 1936.

95. Greenhill, J. P.   Sex and pregnancy. *Redbook* 124:28–30, April, 1965.

96. Guest, M. (Ed.)   *Conference on Biology of the Sperm and of the Vagina.* New York: National Committee on Maternal Health, Inc., 1940.

97. Guze, H.   Anatomy and Physiology of Sex. In *Encyclopedia of Sexual Behavior* (A. Ellis and A. Abarbanel, Eds.). New York: Hawthorn Books, 1961.

98. Haire, N.   *Everyday Sex Problems.* London: Frederick Muller, 1948.

99. Haire, N. (Ed.)   *Encyclopaedia of Sexual Knowledge.* New York: Eugenics Pub. Co., 1937.

100. Hamilton, G. V.   *A Research in Marriage.* New York: Albert and Charles Boni, 1929.

101. Hampson, J. L., and Hampson, J. G.   The Ontogenesis of Sexual Behavior in Man. In *Sex and Internal Secretions* (W. C. Young, Ed.). Baltimore: Williams & Wilkins, 1961.

102. Hardenbergh, E. W.   The psychology of feminine sex experience. *Int. J. Sexol.* 2:224–228, 1949.

103. Hartman, C. G.   *Science and the Safe Period.* Baltimore: Williams & Wilkins, 1964.

104. Hastings, D. W.   *Impotence and Frigidity.* Boston: Little, Brown, 1963.

105. Heiman, M.   Sexual response in women. *J. Amer. Psychoanal. Ass.* 11:360–384, 1963.

106. Heller, C. G., and Maddock, W. O.   The clinical use of testosterone in the male. *Vitamins Hormones* 5:393–423, 1947.

107. Hellman, L. M., *et al.*   Characteristics of the gradients of uterine contractility during the first stage of true labor. *Bull. Hopkins Hosp.* 86:234–248, 1950.

108. Henderson, V. E., and Roepke, M. H.   On the mechanism of erection. *Amer. J. Physiol.* 106:441–448, 1933.

109. Henry, J., and Henry, Z.   Doll Play of Pilaga Indian Children. In *Personality in Nature, Society, and Culture* (C. Kluckhohn *et al.*, Eds.). New York: Knopf, 1953.

110. Herman, M. Role of somesthetic stimuli in the development of sexual excitation in man. *Arch. Neurol. Psychiat.* 64:42–56, 1950.

111. Hess, E. H. Imprinting. *Science* 130:133–141, 1959.

112. Hirsch, J. Individual Differences in Behavior and Their Genetic Basis. In *Roots of Behavior* (E. L. Bliss, Ed.). New York: Paul B. Hoeber, Inc., 1962.

113. Hirschfeld, M. Kastration bei Sittlichkeitzverbrechern. *Z. Sexualwiss.* 15:54–55, 1928.

114. Hirschfeld, M. *Sex in Human Relationships* (J. Rodker, Trans.). London: John Lane, 1935.

115. Hitschmann, E., and Bergler, E. *Frigidity in Women: Its Characteristics and Treatment* (P. L. Weil, Trans.). Washington, D. C.: Nervous and Mental Disease Pub. Co., 1936.

116. Hitschmann, E., and Bergler, E. Frigidity in Women: Restatement and renewed experience. *Psychoanal. Rev.* 36:45–53, 1949.

117. Hohman, L. B., and Schaffner, B. The sex lives of unmarried men. *Amer. J. Sociol.* 52:501–507, 1947.

118. Hollender, M. H. Women's fantasies during sexual intercourse. *Arch. Gen. Psychiat.* 8:86–90, 1963.

119. Hornstein, F. X., and Faller, A. Gesundes Geschlechts Leben. In *Handbuch für Ehefragen.* Olten, Switzerland: Otto Walter, 1950.

120. Hotchkiss, R. S., et al. Artificial insemination with semen recovered from the bladder. *Fertil. Steril.* 6:37–42, 1955.

121. Huhner, M. *Sterility in the Male and Female and Its Treatment.* New York: Rebman, 1913.

122. Huhner, M. The diagnosis of sterility in the male and female. *Amer. J. Obstet. Gynec.* 8:63–75, 1924.

123. Huhner, M. The Huhner test as a diagnosis of sterility due to necrospermia. *Jap. J. Obstet. Gynec.* 19:508–511, 1936.

124. Huhner, M. The Huhner test as a diagnosis of sterility due to necrospermia. *J. Obstet. Gynaec. Brit. Emp.* 44:334–336, 1937.

125. Hunt, M. M. *The Natural History of Love.* New York: Knopf, 1959.

126. Hutton, I. E. *The Sex Technique in Marriage.* New York: Emerson Books, 1931.

127. Javert, C. T. *Spontaneous and Habitual Abortion.* New York: Blakiston Division, McGraw-Hill, 1957.

128. Javert, C. T. Role of the patient's activities in the occurrence of spontaneous abortion. *Fertil. Steril.* 11:550–558, 1960.

129. Johnson, V. E., and Masters, W. H. Intravaginal contraceptive study: Phase I. Anatomy. *Western J. Surg.* 70:202–207, 1963.

130. Johnson, V. E., and Masters, W. H.   Intravaginal contraceptive study: Phase II. Physiology (a direct test for protective potential). *Western J. Surg.* 71:144–153, 1963.

131. Johnson, V. E., and Masters, W. H.   Unpublished data.

132. Johnson, V. E., and Masters, W. H.   A team approach to the rapid diagnosis and treatment of sexual incompatibility. *Pacific Med. Surg.* 72:371–375, 1964.

133. Johnson, V. E., and Masters, W. H.   Sexual Incompatibility: Diagnosis and Treatment. In *Human Reproduction and Sexual Behavior* (C. W. Lloyd, Ed.). Philadelphia: Lea & Febiger, 1964.

134. Johnson, V. E., and Masters, W. H.   A product of dual import: Intravaginal infection control and conception control. *Pacific Med. Surg.* 73:267–271, 1965.

135. Johnson, V. E., Masters, W. H., and Lewis, K. C.   The Physiology of Intravaginal Contraceptive Failure. In *Manual of Contraceptive Practice* (M. S. Calderone, Ed.). Baltimore: Williams & Wilkins, 1964.

136. Kahn, F.   *Our Sex Life* (G. Rosen, Trans.). New York: Knopf, 1939.

137. Kardiner, A.   *Sex and Morality.* Indianapolis, Ind.: Bobbs-Merrill, 1954.

138. Kegel, A. H.   The physiologic treatment of poor tone and function of the genital muscles and of urinary stress incontinence. *Western J. Surg.* 57:527–535, 1949.

139. Kegel, A. H.   Sexual functions of the pubococcygeus muscle. *Western J. Surg.* 60:521–524, 1952.

140. Kelly, G. L.   Technique of Marriage Relations. In *Successful Marriage* (M. Fishbein and E. W. Burgess, Eds.). New York: Doubleday, 1947.

141. Kinsey, A. C.   Sex behavior in the human animal. *Ann. N.Y. Acad. Sci.* 47:635–637, 1947.

142. Kinsey, A. C., *et al.*   *Sexual Behavior in the Human Male.* Philadelphia: W. B. Saunders, 1948.

143. Kinsey, A. C., *et al.*   Concepts of Normality and Abnormality in Sexual Behavior. In *Psychosexual Development in Health and Disease* (P. H. Hoch and J. Zubin, Eds.). New York: Grune & Stratton, 1949.

144. Kinsey, A. C., *et al.*   *Sexual Behavior in the Human Female.* Philadelphia: W. B. Saunders, 1953.

145. Kirkendall, L. A.   Sex Drive. In *Encyclopedia of Sexual Behavior* (A. Ellis and A. Abarbanel, Eds.). New York: Hawthorn Books, 1961.

146. Kirkendall, L. A., and Ogg, E. *Sex and Our Society* (M. S. Stewart, Ed.). Public Affairs Pamphlet No. 366. New York: Public Affairs Committee, Inc., 1964.

147. Kisch, E. H. *The Sexual Life of Women in Its Physiological, Pathological and Hygienic Aspects* (N. E. Paul, Trans.). New York: Allied Book Co., 1926.

148. Kiss, F. Anatomisch-histologische Untersuchungen uber die Erektion. *Z. Anat. Entwicklungsgesch.* 61:455–521, 1921.

149. Kistner, R. W. *Gynecology, Principles and Practice.* Chicago: Year Book Medical Publishers, 1964.

150. Kleegman, S. J. Frigidity in women. *Quart. Rev. Surg. Obstet. Gynec.* 16:243–248, 1959.

151. Klumbies, G., and Kleinsorge, H. Das Herz in Orgasmus. *Med. Klin.* 45:952–958, 1950.

152. Klumbies, G., and Kleinsorge, H. Circulatory dangers and prophylaxis during orgasm. *Int. J. Sexol.* 4:61–66, 1950.

153. Knight, R. P. Functional disorders in the sexual life of women. *Bull. Menninger Clin.* 7:25–35, 1943.

154. Koran The Holy Qur-án (3rd ed.) (Muhammad Ali Maulvi, Trans. and Ed.). Lahore, India: Ahmadiyya Anjuman-i-Isháat-i-Islam, 1935.

155. Krafft-Ebing, R. von *Psychopathia Sexualis: A Medicoforensic Study* (F. J. Rebman, Trans.). Brooklyn, N. Y.: Physicians & Surgeons Book Co., 1922.

156. Krantz, K. E. Innervation of the human vulva and vagina. *Obstet. Gynec.* 12:382–396, 1958.

157. Krause, W. *Handbuch der menschlichen Anatomie.* Hannover, Germany: Hahn, 1841.

158. Krause, W. Die Anatomie des Kaninchens. In *Topographischer und Operative Ruchsicht.* Leipzig, Germany: Engelmann, 1868.

159. Kroger, W. S., and Freed, S. C. Psychosomatic aspects of frigidity. *J.A.M.A.* 143:526–532, 1950.

160. Kubie, L. S. Instincts and homeostasis. *Psychosom. Med.* 10:15–30, 1948.

161. Kubie, L. S. Influence of symbolic processes on the role of instincts in human behavior. *Psychosom. Med.* 18:189–208, 1956.

162. Kuntz, A. *The Autonomic Nervous System* (4th ed.). Philadelphia: Lea & Febiger, 1953.

163. Landis, C., and Bolles, M. M. *Personality and Sexuality of Physically Handicapped Women.* New York: Paul B. Hoeber, Inc., 1942.

164. Lang, W. R. Vaginal acidity and pH: A review. *Obstet. Gynec. Survey* 10:546–555, 1955.

165. Lang, W. R., *et al.* Midvaginal pH. *Obstet. Gynec.* 7:378–381, 1956.

166. Lashley, K. S. Experimental analysis of instinctive behavior. *Psychol. Rev.* 45:445–471, 1938.

167. Lazarus, A. E. The treatment of chronic frigidity by systematic desensitization. *J. Nerv. Ment. Dis.* 136:272–278, 1963.

168. Learmonth, J. R. A contribution to the neurophysiology of the urinary bladder. *Brain* 54:147–176, 1931.

169. Levine, L. A criterion for orgasm in the female. *Marriage Hygiene* 1:173–174, 1948.

170. Levy, D. M. *Maternal Overprotection.* New York: Columbia University Press, 1943.

171. Lewisohn, R. *A History of Sexual Customs.* New York: Harper, 1958.

172. Lichtenstein, H. Identity and sexuality: A study of their interrelationship in man. *J. Amer. Psychoanal. Ass.* 9:179–260, 1961.

173. Lief, H. I. Orientation of Future Physicians in Psychosexual Attitudes. In *Manual of Contraceptive Practice* (M. S. Calderone, Ed.). Baltimore: Williams & Wilkins, 1964.

174. Lief, H. I., *et al.* A psychodynamic study of medical students and their adaptational problems. *J. Med. Educ.* 35:696, 1960.

175. Lloyd, C. W. Sexual Response: Part I. General Considerations. In *Human Reproduction and Sexual Behavior* (C. W. Lloyd, Ed.). Philadelphia: Lea & Febiger, 1964.

176. Loeb, H. Harnrohrencapacitat und Tripperspritzen. *Munchen. Med. Wschr.* 46:1016, 1899.

177. Lorand, S. Contribution to the problem of vaginal orgasm. *Int. J. Psychoanal.* 20:432–438, 1939.

178. Lorand, S. Unsuccessful sex adjustment in marriage. *Amer. J. Psychiat.* 19:1413–1427, 1940.

179. Lyons, W. R., and Templeton, H. R. Intravaginal assay of urinary estrin, *Proc. Soc. Exp. Biol. Med.* 33:587–589, 1936.

180. McClelland, D. C. Studies in serial verbal discrimination learning: III. The influence of difficulty on reminiscence in responses to right and wrong words. *J. Exp. Psychol.* 32:235–246, 1943.

181. McCurdy, H. G. *The Personal World.* New York: Harcourt, Brace & World, 1961. P. 50.

182. McDougall, W. *An Introduction to Social Psychology.* London: Methuen, 1908.

183. McIndoe, A. H., and Banister, J. B. Operation for the cure of

congenital absence of the vagina. *J. Obstet. Gynaec. Brit. Emp.* 45:490–494, 1938.

184. MacLeod, J. Semen Quality in Fertile Marriage. In *Studies on Testes and Ovary, Eggs and Sperm* (E. T. Engle, Ed.). Springfield, Ill.: Charles C Thomas, 1950.

185. MacLeod, J., and Gold, R. Z. The male factor in fertility and infertility: III. An analysis of motile action in the spermatozoa of 1000 fertile men and 1000 men in infertile marriage. *Fertil. Steril.* 2:187–204, 1951.

186. MacLeod, J., and Gold, R. Z. The male factor in fertility and infertility: IV. Sperm morphology in fertile and infertile marriage. *Fertil. Steril.* 2:394–414, 1951.

187. MacLeod, J., and Gold, R. Z. The male factor in fertility and infertility: V. Effect of continence on semen quality. *Fertil. Steril.* 3:297–315, 1952.

188. MacLeod, J., and Gold, R. Z. The male factor in fertility and infertility: VI. Semen quality and certain other factors in relation to ease of conception. *Fertil. Steril.* 4:10–33, 1953.

189. MacLeod, J., and Gold, R. Z. The male factor in fertility and infertility: VII. Semen quality in relation to age and sexual activity. *Fertil. Steril.* 4:194–209, 1953.

190. MacLeod, J., et al. Correlation of the male and female factors in human fertility. *Fertil. Steril.* 6:112–143, 1955.

191. Malchow, C. W. *The Sexual Life.* St. Louis, Mo.: C. V. Mosby, 1923.

192. Malinowski, B. *The Sexual Life of Savages in Northwestern Melanesia.* New York: Halcyon House, 1929.

193. Malleson, J. A criterion for orgasm in the female. *Marriage Hygiene* 1:174, 1948.

194. Marmor, J. Some considerations concerning orgasm in the female. *Psychosom. Med.* 16:240–245, 1954.

195. Maslow, A. H. Self-esteem (dominance feeling) and sexuality in women. *J. Soc. Psychol.* 16:259–294, 1942.

196. Maslow, A. H. *Motivation and Personality.* New York: Harper, 1954.

197. Maslow, A. H. Self-Actualizing People: A Study of Psychological Health. In *Self: Explorations in Personal Growth* (C. E. Moustakas, Ed.). New York: Harper, 1956.

198. Maslow, A. H., et al. Some parallels between sexual and dominance behavior of infra-human primates and the fantasies of patients in psychotherapy. *J. Nerv. Ment. Dis.* 131:202–212, 1960.

199. Masserman, J. H., and Siever, P. Dominance, neurosis, and aggression. *Psychosom. Med.* 6:7–16, 1944.

200. Masters, W. H. Long-range sex steroid replacement—Target organ regeneration. *J. Geront.* 8:33–39, 1953.

201. Masters, W. H. Sex steroid influence on the aging process. *Amer. J. Obstet. Gynec.* 74:733–746, 1957.

202. Masters, W. H. The sexual response cycle of the human female: II. Vaginal lubrication. *Ann. N.Y. Acad. Sci.* 83:301–317, 1959.

203. Masters, W. H. The sexual response cycle of the human female: I. Gross anatomic considerations. *Western J. Surg.* 68:57–72, 1960.

204. Masters, W. H., and Ballew, J. W. The Third Sex. In *Problems of the Middle-Aged* (C. G. Vedder, Ed.). Springfield, Ill.: Charles C Thomas, 1965.

205. Masters, W. H., and Johnson, V. E. The human female: Anatomy of sexual response. *Minnesota Med.* 43:31–36, 1960.

206. Masters, W. H., and Johnson, V. E. Vaginal pH: The influence of the male ejaculate. In *Report of the Thirty-Fifth Ross Conference, Endocrine Dysfunction and Infertility.* Columbus, Ohio: Ross Laboratories, 1960.

207. Masters, W. H., and Johnson, V. E. Orgasm, Anatomy of the Female. In *Encyclopedia of Sexual Behavior* (A. Ellis and A. Abarbanel, Eds.). New York: Hawthorn Books, 1961.

208. Masters, W. H., and Johnson, V. E. The physiology of the vaginal reproductive function. *Western J. Surg.* 69:105–120, 1961.

209. Masters, W. H., and Johnson, V. E. The artificial vagina: Anatomic, physiologic, psychosexual function. *Western J. Surg.* 69: 192–212, 1961.

210. Masters, W. H., and Johnson, V. E. Treatment of the sexually incompatible family unit. *Minnesota Med.* 44:466–471, 1961.

211. Masters, W. H., and Johnson, V. E. Intravaginal environment: I. A lethal factor. *Fertil. Steril.* 12:560–580, 1961.

212. Masters, W. H., and Johnson, V. E. The sexual response cycle of the human female: III. The clitoris: Anatomic and clinical considerations. *Western J. Surg.* 70:248–257, 1962.

213. Masters, W. H., and Johnson, V. E. The sexual response of the human male: I. Gross anatomic considerations. *Western J. Surg.* 71:85–95, 1963.

214. Masters, W. H., and Johnson, V. E. The Clitoris: An Anatomic Baseline for Behavioral Investigation. In *Determinants of Human Sexual Behavior* (G. W. Winokur, Ed.). Springfield, Ill.: Charles C Thomas, 1963.

215. Masters, W. H., and Johnson, V. E. Sexual Response: Part II. Anatomy and Physiology. In *Human Reproduction and Sexual Behavior* (C. W. Lloyd, Ed.). Philadelphia: Lea & Febiger, 1964.

216. Masters, W. H., and Johnson, V. E. Counseling with Sexually Incompatible Marriage Partners. In *Counseling in Marital and Sexual Problems* (A *Physician's Handbook*) (R. H. Klemer, Ed.). Baltimore: Williams & Wilkins, 1965.

217. Masters, W. H., and Johnson, V. E. The Sexual Response Cycles of the Human Male and Female: Comparative Anatomy and Physiology. In *Sex and Behavior* (F. A. Beach, Ed.). New York: John Wiley & Sons, 1965.

218. Mathews, C. S., and Buxton, C. L. Bacteriology of the cervix in cases of infertility. *Fertil. Steril.* 2:45–52, 1951.

219. Maximov, A. A., and Bloom, W. A *Text Book of Histology*. Philadelphia: W. B. Saunders, 1930.

220. Mendelsohn, M. Ist das Radfahren als eine gesundheitsgemässe Uebung anzusehen und aus ärtzlichen Gesichtpunkten zu empfehlen? *Deutsch. Med. Wschr.* 22:381–384, 1896.

221. Menninger, K. A. Impotence and frigidity. *Bull. Menninger Clin.* 1:251–260, 1937.

222. Mering, O., and Weniger, F. L. Social-Cultural Background of the Aging Individual. In *Handbook of Aging and the Individual, Psychological and Biological Aspects* (J. E. Birren, Ed.). Chicago: University of Chicago Press, 1959.

223. Miller, N. F., *et al.* The surgical correction of congenital aplasia of the vagina: An evaluation of operative procedures, end-result and functional activity of the transplanted epithelium. *Amer. J. Obstet. Gynec.* 50:735–747, 1945.

224. Mitsuya, H. Studies on the secretory function of human accessory organs of reproduction. *Jap. J. Urol.* 45:290, 1954.

225. Mitsuya, H., *et al.* Application of x-ray cinematography in urology: Mechanism of ejaculation. *J. Urol.* 83:86–92, 1960.

226. Moench, G. L. Evaluation of motility of spermatozoa. *J.A.M.A.* 94:478–480, 1930.

227. Moench, G. L. A consideration of some aspects of sterility. *Amer. J. Obstet. Gynec.* 13:334–345, 1927.

228. Moll, A. *The Sexual Life of the Child*. New York: Macmillan, 1912.

229. Money, J. Components of Eroticism in Man: Cognitional Rehearsals. In *Recent Advances in Biological Psychiatry* (J. Wortis, Ed.). New York: Grune & Stratton, 1960.

230. Money, J. Components of eroticism in man: I. The hormones in relation to sexual morphology and sexual desire. *J. Nerv. Ment. Dis.* 132:239–248, 1961.

231. Money, J. Components of eroticism in man: II. The orgasm and genital somesthesia. *J. Nerv. Ment. Dis.* 132:289–297, 1961.

232. Money, J. Sex Hormones and Other Variables in Human Eroticism. In *Sex and Internal Secretions* (W. C. Young, Ed.). Baltimore: Williams & Wilkins, 1961.

233. Money, J., et al. Imprinting and the establishment of gender role. *Arch. Neurol. Psychiat.* 77:333–336, 1957.

234. Moraglia, G. B. *Die Onanie biem normalen Weibe und bei den Prostituierten.* Berlin: Priber & Lammers, 1897.

235. Mudd, E. H., et al. Paired reports of sexual behavior of husbands and wives in conflicted marriages. *Compr. Psychiat.* 2:149–156, 1961.

236. Muschat, M. The effect of variation of hydrogen-ion concentration on the motility of human spermatozoa. *Surg. Gynec. Obstet.* 42:778–781, 1926.

237. Muschat, M., and Randall, A. Hydrogen-ion studies on various secretions of the urogenital apparatus. *J. Urol.* 16:515–528, 1926.

238. Myers, R. Evidence of a locus of the neural mechanisms for libido and penile potency in the septo-fornico-hypothalamic region of the human brain. *Trans. Amer. Neurol. Ass.* 86:81–86, 1961.

239. Negri, V. *Psychoanalysis of Sexual Life.* Los Angeles: Western Institute of Psychoanalysis, 1949.

240. Newman, G., and Nichols, C. R. Sexual activities and attitudes in older persons. *J.A.M.A.* 173:33–35, 1960.

241. Newton, N. *Maternal Emotions.* New York: Paul B. Hoeber, Inc., 1955.

242. Nizer, L. *My Life in Court.* New York: Doubleday, 1961.

243. Oberst, F. W., and Plass, E. D. The hydrogen-ion concentration of human vaginal discharge. *Amer. J. Obstet. Gynec.* 32:22–35, 1936.

244. O'Hare, H. Vaginal versus clitoral orgasm. *Int. J. Sexol.* 4:243–244, 1951.

245. Oliven, J. F. *Sexual Hygiene and Pathology.* Philadelphia: Lippincott, 1955.

246. Perloff, W. H. Role of the hormones in human sexuality. *Psychosom. Med.* 11:133–139, 1949.

247. Piaget, G. *The Construction of Reality in the Child.* New York: Basic Books, 1954.

248. Piersol, G. A. *Human Anatomy.* Philadelphia: Lippincott, 1907.

249. Ploss, H. H., *et al.* *Woman*, Vol. I. London: Heinemann, 1935. Pp. 308–340.

250. Potter, E. L. Truths and untruths about miscarriage. *Reader's Digest* 75:77–80, Aug., 1959.

251. Prince, M. *The Unconscious*. New York: Macmillan, 1914.

252. Rado, S. An Adaptational View of Sexual Behavior. In *Psychosexual Development in Health and Disease* (P. H. Hoch and J. Zubin, Eds.). New York: Grune & Stratton, 1949.

253. Rakoff, A. E. Present Problems Related to the Biology of the Vagina. In *Proceedings of the Conference on the Biology of Female Fertility, 1944.*

254. Rakoff, A. E., *et al.* The biologic characteristics of the normal vagina. *Amer. J. Obstet. Gynec.* 47:467–494, 1944.

255. Ramsey, G. V. The sexual development of boys. *Amer. J. Psychol.* 56:217–234, 1943.

256. Rasmussen, J., and Albrechtsen, O. K. Fibrinolytic activity in human seminal plasma. *Fertil. Steril.* 11:264–277, 1960.

257. Reich, W. *The Discovery of the Orgone.* Vol. I: *The Function of the Orgasm.* New York: Orgone Inst. Press, 1942.

258. Reich, W. *The Sexual Revolution.* New York: Orgone Inst. Press, 1945.

259. Reik, T. *Psychology of Sex Relations.* New York: Rinehart, 1945.

260. Retief, P. J. M. Physiology of micturition and ejaculation. *S. Afr. Med. J.* 24:509–514, 1950.

261. Reynolds, S. R. M. *Physiology of the Uterus.* New York: Paul B. Hoeber, Inc., 1939.

262. Reynolds, S. R. M. *Physiological Basis of Gynecology and Obstetrics.* Springfield, Ill.: Charles C Thomas, 1952.

263. Rheingold, J. C. *The Fear of Being a Woman: A Theory of Maternal Destructiveness.* New York: Grune & Stratton, 1964.

264. Riley, F. J., and Masters, W. H. Problems of male fertility: III. Bacteriology of human semen. *Fertil. Steril.* 7:128–132, 1956.

265. Rock, J. Personal communication, 1963.

266. Roheim, G. Women and their life in Central Australia. *J. Roy. Anthropol. Inst. Great Britain & Ireland* 63:207–265, 1933.

267. Rohleder, H. *Die Masturbation: Eine Monographie für Ärzte, Pädagogen und gebildete Eltern* (2nd ed.). Berlin: Fischers medizinische Buchhandlung, 1902.

268. Rohleder, H. *Voresungen über Geschlechstrieb und gesamtes Geschlechtleben des Menchen.* Berlin: Fischers medizinische Buchhandlung, 1907.

269. Rohleder, H. *Die Masturbation: Eine Monographie für Ärzte, Pädagogen und gebildete Eltern* (4th ed.). Berlin: Fischers medizinische Buchhandlung, 1921.

270. Rosenzweig, S. Psychology of the menstrual cycle. *J. Clin. Endocr.* 3:296–300, 1943.

271. Roubaud, F. *Traité de l'Impuissance et de la Stérilité Chez l'Homme et Chez la Femme.* Paris: Baillière, 1876.

272. Rozin, S. Study of seminal plasma: I. The role of seminal plasma in motility of spermatozoa. *Fertil. Steril.* 11:278–285, 1960.

273. Rubin, A., and Babbott, D. Impotence and diabetes mellitus. *J.A.M.A.* 168:498–500, 1958.

274. Rubin, I. C. *Uterotubal Insufflation.* St. Louis, Mo.: C. V. Mosby, 1947.

275. Ruch, F. L. *Psychology and Life* (6th ed.). Chicago: Scott, Foresman, 1963. Pp. 31–33.

276. Ruch, T. C., and Fulton, J. F. *Medical Physiology and Biophysics.* Philadelphia: W. B. Saunders, 1960.

277. Rutherford, R. N., et al. Psychometric testings in frigidity and infertility. *Psychosomatics* 1:3–7, 1960.

278. Sadler, W. S., and Sadler, L. K. *Living a Sane Sex Life.* Chicago: Wilcox & Follett, 1944.

279. Saul, L. J. *Emotional Maturity.* Philadelphia: Lippincott, 1947.

280. Schmidt, J. W. *Libido.* Springfield, Ill.: Charles C Thomas, 1960.

281. Schubert, G. Concerning the formation of a new vagina in the case of congenital vaginal malformation. *Surg. Gynec. Obstet.* 19:376–383, 1914.

282. Scott, J. C. Systolic blood-pressure fluctuations with sex, anger, and fear. *J. Comp. Psychol.* 10:97–114, 1930.

283. Seegar-Jones, G. E., et al. Vaginal fungi and their relation to sperm survival. *Amer. J. Obstet. Gynec.* 70:1271–1276, 1955.

284. Segawa, A. Studies on secretory function of male sexual organs and their accessory glands. *Jap. J. Urol.* 48:869, 1957.

285. Semans, J. H., and Langworthy, O. R. Observations on the neurophysiology of sexual function in the male cat. *J. Urol.* 40:836–846, 1938.

286. Séquy, J., and Somonnet, H. Recherche de signes directs d'ovulation chez la femme. *Gynec. Obstet.* (Paris) 28:657–663, 1933.

287. Séquy, J., and Vimeux, J. Contribution a l'étude des stérilités inexpliquées: Étude de l'ascension des spermatozoides dans voies génitales basses de la femme. *Gynec. Obstet.* (Paris) 27:346–358, 1933.

288. Shedlovsky, L. Some acidic properties of contraceptive jellies. *J. Contraception* 2:147–155, 1937.

289. Shedlovsky, L., *et al.* Titrations of human seminal fluid with acids and alkalies and their effects on the survival of sperm motility. *Amer. J. Physiol.* 136:535–541, 1942.

290. Sherfey, M. J. The evolution and nature of female sexuality in relation to psychoanalytic theory. *J. Amer. Psychoanal. Ass.* (Winter issue), 1965.

291. Sherfey, M. J. Personal communication, 1965.

292. Shor, J. Female sexuality: Aspects and prospects. *Psychoanalysis* 2:47–76, 1954.

293. Shorr, E. Problems of Mental Adjustment at the Climacteric. In *Mental Health and Later Maturity*. Suppl. 168, Public Health Reports. Washington, D. C.: U. S. Public Health Service, 1938.

294. Shorr, E. A new technique for staining vaginal smears: III. A single differential stain. *Science* 94:545–546, 1951.

295. Siegler, S. L. *Fertility in Women*. Philadelphia: Lippincott, 1944.

296. Simmel, E. The significance of psychoanalysis for gynecology. *Calif. & Western Med.* 63:169, 1915.

297. Sims, M. H. Sterility and the value of the microscope in its diagnosis and treatment. *Trans. Amer. Gynec. Soc.* 13:291–307, 1888.

298. Sirlin, J. L. Fate of spermatozoa penetrating into the tissues of the fallopian tube. *Nature* (London) 183:1744–1745, 1959.

299. Spalteholz, W. *Hand Atlas of Human Anatomy* (7th ed.) (L. F. Barker, Trans.), Vol. III. Philadelphia: Lippincott, 1933.

300. Stekel, W. *Frigidity in Woman in Relation to Her Love Life*, Vols. I and II. New York: Liveright, 1926.

301. Stephens, W. N. *The Oedipus Complex: Cross-Cultural Evidence*. Glencoe, Ill.: The Free Press of Glencoe, 1962.

302. Stern, K., and Prados, M. Personality studies in menopausal women. *Amer. J. Psychiat.* 103:358–368, 1946.

303. Stokes, W. R. Sexual functioning in the aging male. *Geriatrics* 6:304–308, 1951.

304. Stokes, W. R. Personal communication, 1963.

305. Stone, H. M., and Stone, A. *A Marriage Manual*. New York: Simon & Schuster, 1953.

306. Street, R. *Modern Sex Techniques*. New York: Archer House, 1959.

307. Symonds, C. P. Concussion and Contusion of the Brain and Their Sequelae. In *Injuries of the Skull and Spinal Cord* (S. Brock, Ed.). Baltimore: Williams & Wilkins, 1940.

334   REFERENCES

308. Talmey, B. S. *Neurasthenia Sexualis*. New York: Practitioner's Pub. Co., 1912.

309. Taylor, G. R. *Sex in History*. New York: Vanguard, 1955.

310. Taylor, H. C., Jr. Vascular congestion and hyperemia: Part I. Physiologic basis and history of the concept. *Amer. J. Obstet. Gynec.* 57:211–227, 1949.

311. Taylor, H. C., Jr. Vascular congestion and hyperemia: Part II. The clinical aspects of the congestion-fibrosis syndrome. *Amer. J. Obstet. Gynec.* 57:637–653, 1949.

312. Taylor, H. C., Jr. Vascular congestion and hyperemia: Part III. Etiology and therapy. *Amer. J. Obstet. Gynec.* 57:654–668, 1949.

313. Terman, L. M. *Psychological Factors in Marital Happiness*. New York: McGraw-Hill, 1938.

314. Terman, L. M. Correlates of orgasm adequacy in a group of 556 wives. *J. Psychol.* 32:115–172, 1951.

315. Thompson, C. *Psychoanalysis: Evolution and Development*. New York: Hermitage Press, 1950.

316. Tinklepaugh, O. L. The nature of periods of sex desire in women and their relation to ovulation. *Amer. J. Obstet. Gynec.* 26:335–345, 1933.

317. Trussell, R. E., and MacDougal, R. F. Vaginal acidity in late pregnancy and its relation to vaginal flora. *Amer. J. Obstet. Gynec.* 39:77–81, 1940.

318. Urbach, K. Uber die zeitliche Gefuhlsdifferenz der Geschlechter wahrend der Kohabitation. *Z. Sexualwiss.* 8:124–138, 1921.

319. Van de Velde, T. H. *Ideal Marriage*. New York: Covici-Friede, 1930.

320. Vaughan, E., and Fisher, A. Male sexual behavior induced by intracranial electrical stimulation. *Science* 137:758–759, 1962.

321. Voge, C. I. B. *The Chemistry and Physics of Contraception*. London: Chapman Hall, 1935.

322. Walker, K., and Fletcher, P. *Sex and Society*. Baltimore: Penguin Books, Inc., 1955.

323. Walker, K., and Strauss, E. B. *Sexual Disorders in the Male*. Baltimore: Williams & Wilkins, 1939. Pp. 177–178.

324. Wallin, P. A study of orgasm as a condition of woman's enjoyment of intercourse. *J. Soc. Psychol.* 51:191–198, 1960.

325. Weisman, A. I. *Spermatozoa and Sterility*. New York: Paul B. Hoeber, Inc., 1941.

326. Westman, A. Investigation into Transport of Ovum. In *Studies*

on *Testis and Ovary, Eggs and Sperm* (E. T. Engle, Ed.). Springfield, Ill.: Charles C Thomas, 1950.

327. Wharton, L. R.   A simple method of constructing a vagina: Report of four cases. *Ann. Surg.* 107:842–854, 1938.

328. Wharton, L. R.   Further experiences in construction of the vagina: Report of twelve cases. *Ann. Surg.* 111:1010–1020, 1940.

329. Wiggers, C. J.   *Physiology in Health and Disease.* Philadelphia: Lea & Febiger, 1939.

330. Wright, H.   *Sex Factor in Marriage.* London: Williams & Norgate, 1930.

331. Wright, H.   A Contribution to the Orgasm Problem in Women. In *Sex, Society, and the Individual* (A. P. Pillay and A. Ellis, Eds.). Bombay, India: International Journal of Sexology, 1953.

332. Young, W. C.   Genetic and Psychological Determinants of Sexual Behavior Patterns. In *Hormones, Brain Function, and Behavior* (H. Hoagland, Ed.). New York: Academic Press, 1957. Pp. 75–98.

333. Zweifel, P.   Die Vaginitis emphysematozoa oder Colpohyperplasia cystica nach Winckel. *Arch. Gynaek.* 12:39–52, 1877.

GLOSSARY

ABDUCT To draw extremities from the midline of the body by means of muscular contractions.

ABLATE To remove by cutting.

ABORTION Expulsion of the products of conception by the twelfth week of pregnancy.

ADDUCT To draw extremities toward the midline of the body by means of muscular contractions.

ADHESION A fibrous band or structure by which tissues are abnormally joined.

ADRENAL HYPERPLASIA Abnormal growth of the adrenal gland usually resulting in overproduction of the adrenal hormones.

AEROBIC Living or growing only in the presence of oxygen.

AFFERENT Carrying to or toward a certain region.

AMENORRHEA Absence of the menses.

AMPULLA A flasklike dilatation at the end of a tubular structure.

ANDROGEN A steroid hormone producing masculine characteristics.

ANTECUBITAL FOSSA The triangular hollow in front of the elbow joint.

APPOSITION Contact between adjacent parts or organs.

AREOLA The ring of darkened tissue on the breast surrounding the nipple.

ATROPHY A failure of nutrition resulting in a wasting away or diminution in the size of an organ or part of the body.

AXILLA The armpit.

BARTHOLIN'S GLANDS Two small glands imbedded in the minor labia at the vaginal orifice which produce a mucoid substance that contributes to the lubrication of the fourchette during prolonged coital activity.

337

CARPOPEDAL SPASM A spastic contraction of the striated musculature of the hands and feet.

CATHEXIS The investment of an object or person with special significance.

CC. *Cubic centimeter.* A fluid measure; 5 cc. are approximately equal to one teaspoonful.

CLIMACTERIC The physical and psychologic phenomena that characterize the termination of menstrual function in the woman and reduction in sex-steroid production in both sexes.

CM. *Centimeter.* A measure of distance; 2.5 cm. are approximately equal to one inch.

COLPORRHAPHY The operation of suturing the walls of the vagina.

COLPOSCOPE An instrument for the visual examination of the vagina and cervix.

CORONA GLANDIS The rim surrounding the base of the glans penis.

CORPUS LUTEUM A yellow mass in the ovary formed from the ruptured graafian follicle. It secretes progesterone, whose function is to prepare the uterus for implantation of the fertilized ovum.

CORPUS UTERI That part of the uterus above the cervix; area of implantation for fertilized ova.

COWPER'S GLANDS Two bulbourethral glands in the penis which secrete a mucoid material as part of the seminal fluid. They may be homologues of Bartholin's glands in the female.

CREMASTERIC The muscles which elevate the testes.

CRUS (pl., *crura*) The leg, stalk, or trunk of a structure, as the two legs of the clitoris that separate and join to the pubic arch.

CUL-DE-SAC (OF DOUGLAS) The rectouterine pouch, formed by a fold of the peritoneum.

CYSTOCELE Protrusion of the urinary bladder through the fascia of the anterior vaginal wall; a hernia.

DARTOS See *tunica dartos.*

DESCENSUS In this usage, the descent of the testes into the scrotum, or the descent of the uterus from the false into the true pelvis after the cessation of effective sexual stimulation.

DETUMESCENCE The subsidence of swelling; in this usage, the loss of localized vasocongestion.

DICRYSTICIN A commercial product now discontinued; contained 300,000 units of procaine penicillin G and 100,000 units buffered potassium penicillin/2 cc.

DISTAL Remote or away from the point of origin.

DYSPAREUNIA Coitus that is difficult or painful for a woman.

DYSURIA Painful urination.

EDEMA Swelling due to accumulation of excess fluid in any tissue or in the intercellular spaces.

EFFERENT Conveying outward or away from a structure; the opposite of afferent.

EGO The individual's concept of self.

EMPHYSEMA The swelling or inflammation produced by the presence of air in any body tissue, particularly the lungs. When present, it magnifies hyperventilation of plateau phase and reduces sexual tension.

ENDOMETRIUM The mucous membrane lining the cavity of the uterus.

EPIGASTRIUM The region lying over the stomach; the upper part of the abdomen just beneath the diaphragm.

EPISIOTOMY Incision in the perineum to facilitate the birth of a child.

EROGENOUS AREAS Those parts of the body that, when stimulated, create subjective erotic arousal.

ERYTHEMATOUS RASH Redness of the skin due to vasocongestion; see *sex flush*.

ESTROGEN A steroid hormone producing female characteristics.

ESTRUS A cyclical period of sexual receptivity in the female animal, during which the sex drive is intense.

ETIOLOGY Causation; the study or theory of the causation of a disease or abnormal functioning of life processes.

EXTEROCEPTIVE Receiving stimulation from the external surface field of the receptor organs.

EXTRAGENITAL   Originating outside or lying outside the genital organs.

FALSE PELVIS   Roughly, the part of the pelvis above the hip joint and the iliopectineal line, the lower part being the true pelvis.

FASCIA   A band of tissue that forms an investment for muscles and certain organs of the body.

FELLATIO   Insertion of the penis into the mouth for purposes of sexual gratification.

FORNIX   The upper portion of the vagina.

FOURCHETTE   The fold of mucous membrane connecting the labia minora along the posterior wall at the vaginal outlet.

FRENULUM   A small fold of skin retained on the ventral surface of the penis after circumcision.

FRIGIDITY   A loosely applied term used to express female sexual inadequacy, ranging from the freudian concept (inability to achieve orgasm through coition) to any level of sexual response considered to be unsatisfactory by either the individual female or her partner on any particular occasion.

FUNDUS   The base of the internal surface of a hollow organ, as the fundus of the uterus.

GLUTEAL   Pertaining to the buttocks.

GLYCOGEN   A carbohydrate form of food material stored in the liver, muscles, and some other tissues.

GRAVIDA   A pregnant woman; referred to as gravida I, or primigravida, during the first pregnancy, gravida II during the second pregnancy, etc.

HERNIORRHAPHY   The operation of suturing a hernia.

HETEROSEXUALITY   Sexual interest in or sexual activity between members of the opposite sex.

Hg   Chemical symbol for the element mercury.

HOMOLOGOUS   Having a corresponding position, structure, and origin with another anatomical entity; as, an organ or part of one sex being comparable to a unit in the opposite sex. These organs may or may not have the same function.

HOMOSEXUALITY Sexual interest in or sexual activity between members of the same sex.

HUMORAL Pertaining to any fluid of the body.

HYPERTROPHY The excessive enlargement or overdevelopment of an organ or part; the opposite of *atrophy*.

HYPERVENTILATION Excessively rapid and deep breathing.

HYSTERECTOMY Surgical removal of the uterus, either through the abdominal wall or through the vagina.

IMPOTENCE Disturbance of sexual function in the male that precludes satisfactory coitus. It varies from inability to attain or maintain full erection to total loss of erective prowess. *Primary impotence*: difficulty from the onset of sexual activities. *Secondary impotence*: difficulty which arises later in life, following a history of effective sexual functioning.

INGUINAL Pertaining to the groin.

INTEGUMENT The outer covering, especially the skin.

INTERCOSTAL Between the ribs.

INTIMA The innermost of the three coats of an artery.

INTROITUS The entrance to a cavity or space, e.g., the vagina.

INTROMISSION Insertion of the penis into the vagina.

IN VITRO Observable or occurring outside of the living organism.

IN VIVO Within a living organism.

INVOLUNTARY Performed independently of the will.

INVOLUTION Retrograde development; a decline of physical or mental function.

ISCHIUM The inferior dorsal portion of the hip bone.

"LETHAL FACTOR" A factor within the vaginal environment of some women that immobilizes spermatozoa within seconds in severe concentrations and within minutes in more dilute concentration.

LIBIDO   Sexual drive or urge.

LUMEN   The internal cavity or interior of a tube.

MACROSCOPIC   Large enough to be observed by the naked eye, as opposed to *microscopic*.

MACULOPAPULAR   Spotted and raised or elevated.

MASTURBATION   Self-stimulation of the sexual organs.

MEATUS   An opening, such as the end of the urethral passage through the penis.

MENOPAUSE   The period of cessation of menstruation in the human female, occurring usually between the ages of 45 and 50.

MENSURATION   The process of measuring.

MICTURITION   The act of urinating.

MISCARRIAGE   Expulsion of a fetus from the onset of the fourth to the end of the sixth month of pregnancy.

MONOGAMOUS   Pertaining to monogamy, or marriage to but one person at a time.

MOUNT   To make the initial thrust of the penis into the vagina with onset of coition.

MUCOID   Resembling mucus.

MUCOSA   A mucous membrane; a thin tissue that has a moist surface.

MULTIPARA (adj., *multiparous*)   A woman who has given birth to two or more children.

MYOMA   A tumor consisting of fibrous and muscle tissue that grows in the wall of the uterus: also called *fibroid*.

MYOMETRIUM   The muscular substance of the uterus.

MYOTONIA   Increased muscular tension; a secondary physiologic response to sexual stimulation.

NEONATE   A newborn infant.

NULLIPARA (adj., *nulliparous*)   A woman who has never borne a viable child.

OOPHORECTOMY   Surgical removal of an ovary.

OPISTHOTONOS   A form of tetanic spasm in which the head is bent backward and the body is bowed forward.

OS   Mouth or orifice, as the os of the cervix.

OVULATION   The release of an ovum from the graafian follicle of the ovary.

PARITY   The condition of having borne a child or children: Para 1, one child; para 4, four children; etc.

PATHOGNOMONIC   Characteristic of a certain disease; used to describe signs or symptoms by which a diagnosis can be made.

PATULOUS   Open, expanded; spread widely apart.

PERINEUM   The area between the thighs, extending from the posterior wall of the vagina to the anus in the female and from the scrotum to the anus in the male.

PERITONEUM   The strong, transparent membrane lining the abdominal cavity.

pH   The symbol used in expressing hydrogen ion concentration, the measure of alkalinity and acidity. pH values run from 0 to 14, 7 indicating neutrality, numbers less than 7 increasing acidity, and numbers greater than 7 increasing alkalinity.

PHALLUS   The penis.

PHIMOSIS   Tightness of the foreskin of the penis so that it cannot be drawn back over the glans.

POSTPARTUM   Occurring after delivery.

PRIMIGRAVIDA   A woman who is pregnant for the first time; also written Gravida I.

PRIMIPARA (adj., *primiparous*)   A woman who has borne but one child.

PROGESTERONE   See *corpus luteum*.

PROPHYLAXIS   Preventive treatment.

PROPRIOCEPTIVE   Receiving stimulation within the tissues of the body.

PROTEAN   Variable; readily assuming different shapes or forms.

PSYCHE   The thinking and emotional faculty in man, including both the conscious and unconscious processes.

PSYCHOGENIC   Of psychic origin.

PUBIS   The anterior inferior part of the hip bone.

PUDENDUM   The mons pubis, labia majora, labia minora, and the vestibule of the vagina.

RECEPTOR   A sensory nerve terminal that responds to stimuli.

RECTOCELE   Protrusion of part of the rectum into the posterior floor of the vagina; a hernia.

REFLEX ARC   A neural mechanism. Any action that takes place through such a mechanism is a relatively simple, automatic response to a stimulus which is independent of the higher nerve centers of the brain.

REFRACTORY PERIOD   A temporary state of psychophysiologic resistance to sexual stimulation immediately following an orgasmic experience.

RETRACTION REACTION   The retraction, during plateau phase, of the clitoral body from its normal pudendal-overhang positioning.

RETROGRADE   Directed backward; reverse.

RUGAL   Wrinkled, corrugated.

SALPINGECTOMY   Surgical removal of an oviduct (fallopian tube).

SEX FLUSH   The superficial vasocongestive skin response to increasing sexual tensions (plateau phase-oriented).

SEX SKIN   In this usage, the preorgasmic discoloration of the minor labia of the human female (plateau phase-oriented). ·

SEX STEROIDS   A group name for compounds including sex hormones with estrogenic and androgenic properties.

SEXUAL INADEQUACY   Any degree of sexual response that is not sufficient for the isolated demand of the moment; may be constant or transitory inability of performance.

SIMS-HUHNER TEST   Postcoital examination of ovulatory cervical mucus usually conducted within two to three hours after coition.

SOMATOGENIC   Of bodily origin.

SPECULUM (pl., *specula*)   An instrument for the visual examination
of a passage or cavity of the body, e.g., the vagina.

SPHINCTER   A ringlike muscle that surrounds and is able to close a
natural opening.

SUBJECTIVE   Perceived only by the person affected; not perceptible
by another.

SUBSTRATE   An underlying layer or support possessing special
qualities.

SWEATING PHENOMENON   An early phase in the production of vaginal lubrication before coalescence of the material occurs.

SWEATING REACTION   An involuntary perspiratory reaction that occurs during the immediate postorgasmic portion of the resolution
phase in the sexual response cycle of both men and women.

SYNDROME   A group of symptoms which characterize a particular
disorder.

TACHYCARDIA   Excessively rapid heart action.

TENESMUS   Ineffectual and painful straining to defecate or urinate.

TESTOSTERONE   The male testicular hormone. Its function is to
produce and maintain male secondary sex characteristics. Is involved in both protein and carbohydrate metabolism.

TRABECULA (pl., *trabeculae*)   A band or cord of fibrous and smooth
muscle tissue.

TRANSFORMER   In this usage, an organ or a part of one functioning
to step up or to increase reactive energy or potential.

TRANSUDATE   Any substance that has passed, like sweat, through
the pores of tissues.

TUMESCENCE   The process of swelling or enlarging.

TUNICA DARTOS   A layer of smooth-muscle fibers in the superficial
fascia of the scrotum.

URETHROCELE   Protrusion of the female urethra through the fascia
in the anterior vaginal wall; a hernia.

ᴀGINAL AGENESIS Congenital absence of the vagina.

VAGINAL AXIS An imaginary line running directly from the vaginal outlet to the depth of the cul-de-sac.

VAGINAL BARREL The vaginal cavity.

VAGINAL LUBRICATION A transudation-like material appearing on the walls of the vaginal barrel within a few seconds from the onset of any form of effective sexual stimulation. The first physiologic response to sexual stimulation in the female.

VARICOCELE Enlargement of the veins surrounding the spermatic cord in the male.

VARICOSITY A condition or state in which a vein is irregularly swollen.

VASOCONGESTION Congestion of the blood vessels. The primary physiologic response to sexual stimulation; primarily venous blood.

VASODISTENTION Distention and dilation of blood vessels leading to increased blood supply to the part; opposite of vasoconstriction.

VESTIBULE Space or cavity at the opening to a canal, such as the vaginal outlet.

VISCERAL The internal organs.

VISCOSITY The quality or state of being sticky or glutinous.

VOLUNTARY Accomplished under control of the will.

Labia minora, 40–42, 289
  in aging women, 231–232
  color changes in, 7, 41–42, 53,
      184, 231, 281, 289
    in aging women, 231–232
    with artificial vagina, 107, 109
    postpartum, 151, 152
    in pregnancy, 147
  in excitement phase, 280, 289
  expansion of, 40–41, 42, 50, 119,
      122, 289
  in orgasmic phase, 289
  penile traction on, 59, 60
  in plateau phase, 289
  in pregnancy, 146, 147
  in resolution phase, 284, 289
  tactile sensitivity of, 64
Lactation period
  breasts in, 144, 150, 151
  eroticism in, 161, 162
Leg muscles, in plateau phase, 298–
  299
Lethal factor, vaginal, 97–100
Lubrication, vaginal. See Vagina,
  lubrication in

Male sexual response, 5, 171–220
  age range of subjects, 13, 248,
      249
  in aging, 18–19, 248–270
    ages of subjects, 248, 249
    automanipulation, 262
    breasts in, 249–250
    and concern with economic
      pursuits, 265–266
    ejaculation, 257–259
    extragenital reactions, 249–251
    fatigue affecting, 266–267
    fear of failure, 269–270
    frequency of exposure to sexual
      episodes, 251, 262–263, 269
    impotence, 203
    and monotony in sexual rela-
      tionship, 264–265
    myotonia in, 250–251
    noctural emission, 262
    and overindulgence in food and
      drink, 267–268

penis in, 181, 251–254
and physical and mental infirmities, 268–269
psychic influences on sexual responses, 260, 263–267, 268–
  270
rectum in, 251
scrotum in, 254–255, 259
sex flush in, 250, 259
sexual inadequacy in, 203, 263,
  270
testes in, 255–256, 259
vasocongestion in, 259
automanipulation, 197–200, 262,
  313
  See also Automanipulation, in
    males
blood pressure in, 174–176, 291
breasts in, 171–172, 249–250, 290
  See also Breasts, in males
cardiac patients, 175–176
cardiorespiratory reactions, 174–
  176
carpopedal spasm in. See Carpopedal spasm
and circumcision, 17, 18, 189–
  191, 198
Cowper's glands, 211, 281, 293
education among study subjects,
  14, 15, 156, 261
ejaculation. See Ejaculation
and eroticism, 263–270, 301, 302
excitement phase. See Excitement
  phase, male
extragenital reactions, 171–176,
  249–251, 290–291
fears of performance, 200–203,
  218, 269–270
genital reactions, 292–293
homosexuality, 22, 200, 201, 219,
  306, 308, 309
husbands of pregnant women,
  141, 156, 160, 163–165
hyperventilation in, 174, 291
  compared with females, 277–
  278
impotence. See Impotence in
  males

compared with females, 278–279

pH
  of seminal fluid, 99
  vaginal, 89–100
Phallic fallacies, 188, 189, 190, 191–193, 200–203
  and clitoral response, 45, 57
Phimosis, 180, 190
Physical examinations of subjects, 12, 22, 189
Physical infirmities, effects of, 268–269
Plateau phase, 4, 6
  in aging females
    Bartholin's glands in, 232
    breasts in, 225–226
    clitoris in, 230
    uterus in, 237
    vagina in, 235
  in aging males
    penis in, 252
    scrotum in, 255
    testes in, 255–256
  duration of, 6
  in females
    aging females, 225–226, 230, 232, 235, 237
    artificial vagina in, 106–107
    Bartholin's glands in, 43, 44, 232, 289
    blood pressure in, 35, 287
    breasts in, 29–30, 143, 225–226, 274, 286
    carpopedal spasm in, 33
    clitoris in, 51–52, 58, 65–66, 230, 288
    compared with males, 280–282
    hyperventilation in, 34, 277, 287
    labia majora in, 39, 40, 289
    labia minora in, 7, 40, 41, 42, 281
    mucoid emissions in, 281–282, 289
    myotonia in, 32, 276, 286, 294, 295–299
    pelvis in, 76

postpartum, 150, 151, 152
in pregnancy, 147–148
rectum in, 34, 287
sex flush in, 32, 227, 275, 286
tachycardia in, 34, 278, 287
uterus in, 112, 120, 237, 288
vagina in, 75–77, 84–85, 235, 280–281, 288
vasocongestion in, 280–282
in males
  aging males, 252, 255–256
  blood pressure in, 291
  breasts in, 171, 274, 290
  compared with females, 280–282
  Cowper's glands in, 293
  hyperventilation in, 174, 277, 291
  mucoid emissions in, 281, 289
  myotonia in, 173, 276, 290, 294, 295–299
  penis in, 7, 183–184, 192, 252, 280, 281, 292
  preejaculatory fluid emission in, 210–211
  rectum in, 173, 290
  scrotum in, 205, 255, 292
  secondary organs in, 293
  sex flush in, 172, 250, 275, 290
  tachycardia in, 174, 278, 291
  testes in, 207–209, 255–256, 293
  urethra in, 187
  vasocongestion in, 280–282
myotonia in, 7
vasocongestion in, 7
Population for research studies, 9–23
  age of subjects, 12, 13
  artificial vagina group, 12, 15, 18
  clinic sources, 11, 14–15
  education levels of subjects, 11, 13–14, 15
  family units, 12, 306–307, 309–311, 312, 314
  female subjects, 12, 13
  geriatric group, 18–19
    females, 223–224

Tenting effect—*Continued*
in aging women, 237
Testes, 206–209, 293
in aging male, 255–256, 259
elevation of, 206–207, 255, 276, 280, 293
in aging males, 254, 255–256
enlargement of, 203, 293
in excitement phase, 206–207, 255, 293
involution of, postorgasmic, 209
in aging males, 256
in orgasmic phase, 209, 256, 293
in plateau phase, 207–209, 255–256, 293
in resolution phase, 209, 256, 284, 293
vasocongestion in, 208, 293
in aging males, 256
Transformer role
of clitoris, 62–63
of penis, 196–197
Tumescence. *See* Vasocongestion.

Urethra
in females, 33
in aging women, 227–228
contractions of, 130, 283
in males, 187–188
contractions of, 213, 215, 217, 282, 283
Urinary loss, in coition, 228
Urination
painful, in women, 33, 228, 240
penis in, 187
postcoital urge for, 33
Uterus, 111–126, 288
as abdominal organ, in pregnancy, 167
in aging women, 236–238
cervical response, 114–116
in aging women, 236–237
cervical secretory activity, 69, 70–71, 114–115
in aging women, 236
in pregnancy, 147
contractions of, 115, 116–119, 122, 126, 129, 133, 141, 282–283, 288

in aging women, 237–238, 241
from automanipulation, 118, 126, 165
painful, 118–119
in pregnancy, 137, 148
sensation of, after uterine removal, 118
corpus response, 116–122
dilatation of cervical os, 115, 288
in aging women, 237
elevation reaction, 112–114, 124, 288
in aging women, 237
postpartum, 152
in pregnancy, 148
in excitement phase, 73, 82, 112, 288
increased size of, 119–122
involution of, in aging, 237
in menstruation, 124–126
myotonia of, 116–119
in orgasmic phase, 112, 116–119, 122, 126, 288
parity affecting responses of, 115, 119, 120–121, 122, 129
in plateau phase, 112, 120, 288
postcoital cervical mucus examination, 89, 92
postpartum, 151, 152
in pregnancy, 147
in resolution phase, 78–79, 87, 112–113, 115, 288
response to sexual stimuli, 111, 114–116
and sperm migration, 115–116, 122–124
retroversion of, 73, 86
and lack of elevation, 114
in pregnancy, 147
secretory activity of, 111
sucking concept of, 122–124
vasocongestion in, 119–122
in aging women, 237

Vagina, 68–100, 288
acidity of, 44, 89–90
neutralization of, 94
seminal fluid affecting, 98

CPSIA information can be obtained at www.ICGtesting.com
Printed in the USA
LVOW10s0949240713

344281LV00002B/267/P